IET MANAGEMENT OF TECHNOLOGY SERIES 24

The Art of Successful Business Communication

Other volumes in this series:

The Art of Successful Business Communication

Patrick Forsyth

with Frances Kay

The Institution of Engineering and Technology

Published by The Institution of Engineering and Technology, London, United Kingdom

© 2008 The Institution of Engineering and Technology

First published 2008

The Institution of Engineering and Technology
Michael Faraday House
Six Hills Way, Stevenage
Herts, SG1 2AY, United Kingdom

www.theiet.org

British Library Cataloguing in Publication Data
A catalogue record for this product is available from the British Library

ISBN 978-0-86341-907-2

Typeset in India by Newgen Imaging Systems (P) Ltd, Chennai
Printed in the UK by Athenaeum Press Ltd, Gateshead, Tyne & Wear

One should not aim at being possible to understand, but at being impossible to misunderstand.

Quintillian, Roman rhetorician

Contents

Preface

We all communicate, much of the time, and the workplace is no exception. Often, all goes well. Often, we hardly think about it. Indeed, how difficult is it to say, 'What time do you call this?' to the postman or ask for a salary increase, make a presentation to the board or write a report that will actually be read and influence a decision towards the one you want made?

Well, leaving aside the postman, the answer may be not only that such things can be difficult, but also that when they are poorly executed problems are not far behind. In most workplaces you do not have to eavesdrop for long to hear the immediate results of poor communication:

'But I thought you said ...'

'You should have said ...'

'*What!?*'

Similarly, failing to get your point across at a meeting or making a lacklustre presentation can change the course of subsequent events – to your detriment.

There are difficulties: for instance, those making a poor presentation often cite lack of time to prepare as an excuse. More often than some recognizable fault destroying or diluting the effectiveness of communications, it is *lack* of any thought that jeopardizes it. It is assumed all will be well and no great thought or preparation occurs.

This is dangerous, because the fact is that communication is often *not* easy; indeed a host of factors combine to make it more difficult. Here, in a publication made available by the Institution of Engineering and Technology, a further factor is relevant. If 'straightforward' communication can be problematical, how much more so is that the case when technicalities are involved – especially when communication is directed from specialist to layman. As an illustration, think about this: how quickly and easily could you tell someone who doesn't know how to tie a necktie? And, no, you cannot demonstrate – words only!

This book addresses these problems: what makes communication difficult and how to overcome those difficulties; how to deal with specific modes of communication (for instance, making a presentation or putting something in writing); and, overall, it highlights the *opportunities* that good communication produces. Two further things. First, there is a logic to the book, and the content of early chapters may enhance the reading of a subsequent one on, say, negotiation, but chapters on individual methods make sense in their own right; indeed, some points are repeated to ensure this. Second, the book is intended to be useful whatever role someone may have. So, for example, the chapter on meetings addresses both those who attend them and those who run them. Similarly, skipping one chapter that is not relevant to you will not affect the overall continuity. The good communicator can influence things, swing opinions, persuade and build their own image and reputation in the process – communicating is a career skill (influencing how you fare in the workplace) as well as a necessary work one.

There is a need to take it seriously, but, that said, the process is essentially common sense and without a doubt the thinking reading this book can prompt will make it more likely that you will communicate effectively and achieve what you want through your communications.

Patrick Forsyth
Touchstone Training & Consultancy
28 Saltcote Maltings
Maldon
Essex CM9 4QP

Chapter 1
Communication: its nature, scope and purpose

1.1   A common thread

Leaving other specialist business skills on one side for the moment – only because what they are may vary somewhat for people in different roles – consider what links so many aspects of what goes on in the business world. The common thread is, in a word, communication.

> *Communication*: the passing on to another person (or people) of a specific message by whatever method. In a business context it should be assumed that clarity is always intended and that messages have specific purposes.

Almost everything one might list in the same breath as the word *business* involves, or is a form of, communication. The implication is clear: to be effective in business, someone must be a good communicator. To quote Lord Gormley: 'You impress folks that little bit more with what you're saying if you say it nicely. People don't hear your ideas if you just stand there shouting out words.'

It is a truism to say, certainly of executive and managerial roles, that: If you cannot communicate clearly you cannot operate in business effectively. Not all people are good communicators, of course. Some may admit it, but 'muddle through somehow', perhaps feeling other characteristics are more important, or make up for shortfalls in communication. Perhaps an alternative maxim is a better starting point for a book such as this.

> Good communication can make success more certain and make you more likely to achieve excellence in job and career.

And that is the theme of this book. It is not its task to review all the skills involved in business, nor to investigate deeply the technical factors that they may, necessarily, need to incorporate. For example, in dealing with people, such specialist matters as employment legislation and unionization may be occasionally important; specialist knowledge and skills of all sorts may be mentioned or implied here, but the book does not set out to cover such matters in detail. This is not to negate the importance of such areas, but the intention is to follow the common theme, to focus on the specific part communication plays within the business process and review in what ways (a) *what* is done, and (b) *how* it is done can help the process.

1.2   The special ingredient

Most economies around the world have been through difficult, or at least different and unpredictable, times in recent years; indeed 'getting back to normal' does not seem an option for which it is worth waiting. It is axiomatic that people bear the brunt of any commercial or organizational challenge. They must cope with declining markets, ever more fickle customers, financial and corporate upheaval and a more dynamic business environment than has ever existed in the past. The IT (information technology) revolution, to take one example, is evidence enough of the rate of change everyone in business must contend with daily, perhaps especially managers. It is not just doom and gloom that create problems. Positive developments, too – for example, as a company launches a new product onto a growing market – are equally likely to produce challenges. Of course, when success is achieved, many things influence it. Some, such as sheer hard work or persistence, are likely to be as important as more technical or innovative ones.

But when push comes to shove it is *people* who create success. And managers have the job of leading and supervising their people. Take people out of the organization and there is little, if anything, left. Similarly, take the communication out of the operation of a business and the organization ceases to exist in any meaningful way.

If people make a difference, then communications excellence is a part of how things are made to work well. It may not be a real magic formula (if only a real one did exist!), but attention to it can certainly produce improved results.

1.3   The management dimension

Jobs in business vary. Business encompasses those who maintain a situation, who keep a department or section not simply ticking over but performing efficiently and productively, but perhaps with no pressing need to develop or innovate. It also includes those whose job is inherently better described as concerned with the initiation of things, with innovation, change and creativity. Whatever the spectrum involved, and it should be said that coping with or initiating change affects more and more managers, everyone is dependent on other people – on the team that works for them or with which they work. 'Other people' could be secretaries or assistants, a department of twenty people or a whole organization. Now, make no mistake, business is demanding and multifaceted. To be effective, it needs the application of consideration, time and effort.

Good communication cannot change this fact, but it can raise the chances that what is done works well. For just a while, let me focus on the job of those who manage other people (though the book is aimed equally at those who do not do so). For instance, the best managers treat their team like royalty. They are the most important people in their lives. They work at helping them succeed. In simple terms, and assuming they have the ability to do the job, ideally what is required is the creation of a team who work:

- efficiently
- effectively
- reliably
- consistently
- productively.

Ideally it is also necessary for people to be focused on the job and to have an appropriate degree of self-sufficiency (*empowered* was the word that enjoyed a vogue for a while) so that the manager can manage from a distance. If good performance is achieved only by watching people every step of the way, this is time-consuming and hardly reflects effective team performance. A good team is quick on its feet as it were, and that too is a characteristic brought out by good management. Even in other relationships, where direct reporting relationships are not involved, communication can oil the wheels and make things run smoothly.

1.4   The nature of communication

Communication is inherent to the conduct of business and businesses. Whether you want to prompt a specific action, instigate discussion or idea generation, change attitudes or go through specific communications processes such as appraisal, say, it all starts with communication – and good communication can ensure or enhance all these processes. But it may stop there too. So first let us consider the problems poor communication can create.

1.4.1   Negative effects

The list of ways in which poor, or ill-judged, communication might have negative impact is legion. The evidence of poor communication is all around us in most (all?) organizations. Walk through the typical office and you will hear the likes of the following floating in the air:

'I thought you meant…'
'But you said to…'
'No, what I meant to say was…'
'Why didn't you say that in the first place?'

Sometimes such conversations do no great harm, at least not beyond a momentary hiatus while something is sorted out, perhaps in an additional phrase or two. On other occasions more harm is done: incorrect action is taken; time is wasted; money is spent unnecessarily; and deadlines are missed. The effect can be external, too, resulting in upset, inconvenienced or *former* customers, for example. There is a dilution of effectiveness at work here, the dangers of which will be readily clear (we look at exactly why this sort of thing happens and how to avoid it in the next chapter).

All this may come from a brief, but ill-thought-out, few words of conversation. Or much more time and effort may be involved. Someone might, for example, spend hours writing a long, detailed report only to find that it was unnecessary, and that the instructions given had meant something else. This is something that is not just an example of waste and inefficiency, but can be personally demoralizing to those involved as well.

Consider some examples before moving on. The following all show how particular tasks within an organization are affected by communications failure.

- *Recruitment and selection* can be a chore. It is, however, a vital task because having the right people in place is a differentiating factor for any organization. Yet one ill-prepared interview, maybe just a few questions asked in the wrong way (or not asked at all), and the result – the right candidate missed or the wrong one appointed – may produce consequences that must be lived with for a considerable time.
- *Appraisal* is another major interviewing task with similar potential for problems. Again, such a meeting is not the easiest thing to conduct, and if communications break down then maybe an employee who should be nurtured for the future benefit of the firm is found to be leaving in six months' time.
- *Time off*: here is something much smaller-scale. A member of staff asks for time off (let us assume for what is a good reason). Pressure of the moment and looking ahead prompts an offhand, negative answer, and motivation – as well as productivity – is immediately affected and takes time to repair.

The desirability of avoiding such instances as those cited above, large or small, will be clear. But the reverse of all this is perhaps where the focus must lie. It is not enough to avoid breakdowns in communication and get the communication right: it is important to get the most from the situation by executing the communication as well as possible.

1.4.2   Positive impact

The clarity of any message clearly has an effect on what occurs after it is delivered. Potential problems have already been hinted at. At best, poor communication produces confusion; at worst, it fails to get done whatever should be done. Conversely, ensuring that a message is clear and unambiguous can result in positive action. Exactly what should get done gets done. Such communication is directly able to:

- prompt or speed up action;
- improve efficiency;
- increase productivity;
- stimulate creativity.

Indeed it will act as a spur to whatever action is required. This may literally be an action; for instance, some management communication is in the nature of an instruction. But it may also be designed for other purposes, say, to:

- inform;
- instruct;
- motivate;
- change opinions;
- prompt debate or discussion;
- stimulate the generation of ideas;
- build on prior contacts or thinking.

Such a list could doubtless be extended and makes the point that there is a great deal hanging on any communication between management and staff and around the organization that it is worth getting it right if such intentions are to be achieved as you wish. As we will see, communication can be a problem; its very nature can easily produce confusion. Chapter 2 investigates something of the psychology involved, what causes communications breakdowns and what helps prevent them. Here, even a small example makes a point: there is all the difference between asking someone 'to get some information out immediately' (what is immediately, exactly – as soon as possible?) and saying that it must go to someone 'by email before three o'clock this afternoon'. This sort of precision is just one of the ways in which accuracy can be achieved and result in the appropriate outcome being ensured just that much more certainly. Specific examples of how activity can be positively affected by good communications appear below. The following reflect topics other than those mentioned in a negative light above (although each has their positive side).

- *Training* can be very valuable (and I say this not just because I undertake training work!). But this is not always the case: a briefing meeting where time is skimped and needs wrongly identified can result in a member of staff attending a course that they do not enjoy, from which neither they nor the organization gains benefit and which sours the view of training for the future. Good pre-course (and post-course, for that matter) communication can enhance the training experience, changing a planned course attendance from something viewed as an awkward break in other work to something that is looked forward to, worked at open-mindedly and from which someone draws real benefit.
- *Incentives* are designed to prompt additional effort and make targeted results more likely to be achieved. Incentive schemes are not a universal panacea, yet can be very worthwhile in the right context. Yet more than one has failed because managers fail to check or listen and end up instigating a scheme with no appeal to the people it is intended to influence (sometimes perhaps the awards are picked solely because they appeal to the manager!). Discussion beforehand can help devise an appropriate scheme; clear communication of the whys and wherefores of it can ensure it hits the spot and works well.
- *Rumour and bad news* is another danger area. 'Leave it alone, do nothing and it will go away' is sometimes the most tempting attitude to adopt. This is more

because dealing with it is awkward rather than because of a real belief that this will work. However, clear, positive communication powerfully put over, at the right time, can stop a rumour dead in its tracks and get motivation heading in the right direction again.

1.5   The bonus of message plus method

Communication involves three elements: the message, the delivery method used and the messenger. All are important.

1.5.1   The message

The clarity of any message clearly has an effect on the results it is intended to achieve. But the message alone is not the sole influence on how it is received. How we ensure that messages are put across in the right way constitutes much of the content of this book.

1.5.2   The method

The method matters, too. There are things that are best done at a meeting; in a letter, memo, fax or email; one to one; on the telephone; or in a moment as two people pass in a corridor. Yet each method is as unsuitable for some things as it is right for others. Few would appreciate being fired by email. Well, few are likely to appreciate being fired *however* the message is delivered, but the point is that the method can make it worse. One might say the same of the reverse: being promoted surely deserves some discussion and a sense of occasion; that too might seem inappropriately handled if an email, say, was the sole means of communication. Combinations can be used in different ways. A promotion might be discussed, confirmation sent as an email (to delay the good news not at all), discussion might then again follow and a written confirmation – a letter or memo – complete and confirm the detail.

In every circumstance one of the things that deserve thought is the choice of method. This means a concentration on the recipient and the result. It may be quicker and easier just to lift a telephone, but other ways may have more lasting impact and power. Consider how much stronger the effect is of receiving a thank-you letter compared with a quick telephone call.

Of course, it is horses for courses. Every method has its place but each is best for some applications and less good for others.

1.5.3   The messenger

What can the individual bring to bear on all this?

1.5.3.1   The power of positive image

The view people hold of an individual will also have an impact on the way their communications are regarded and indeed acted upon. The reasons why someone is regarded as they are, by their immediate colleagues or staff and others, are not easy to tie down. Many factors are at work here: someone's nature and personality; their competence, expertise and experience; all aspects of their management style; and, not least, how successful they are and what results they achieve. Even their appearance plays a part.

Certainly, an important element is their whole communications style and ability.

A manager, say, who never has time for anyone, especially for consultation, who conducts their relationships through minimal, monosyllabic dialogue and terse, one-line memos, will rarely endear themselves to others. Nor will the person who waffles, never expresses a real opinion nor voices a clear statement.

It may be no fault of the messages themselves, but such signals can dilute the good impression or good influence that such people would otherwise make on others. And this is an aspect of image building that can be compounded by lack of skills in particular areas of communications method. Here is meant such things as the manager who collects a reputation for being ineffective at such things as handling a meeting, making a presentation or writing a report. It is difficult for people to take on board even the most sensible message when it is buried in a dense report notable only for its length, profusion of gobbledegook and 'office-speak' and convoluted structure. Similarly, people are very harsh about certain methods. They are not likely to say, 'What excellent ideas! What a pity they weren't better presented!' They are more apt to say, 'What a rotten presentation! I bet the ideas weren't up to much, either.'

The converse of this is also true. Good communicators are inherently more likely to be held in respect. What they say is, almost automatically, treated with more respect than what others, less adept or careful in this respect, say. Recipients pick up confidence and competence in communicating. They are more likely to pay attention, think about and give real consideration to messages they see as well conceived and well directed, and this will directly help the results of those who are seen in this positive light.

This is an effect that operates actively. In other words, people look at those communicating with them and actively seek to use their style and approach to assist in the judgements they make about the content of messages. If you doubt this happens, think of what occurs when even appearance contributes something very specific to meaning. If a voice says, 'Excuse me' as you walk down the street, then your reaction surely takes immediate account of the fact if you turn to see that a uniformed policeman said it.

The moral here is to act to ensure that you develop and use communications skills in a way that gains this effect, even if it means sometimes operating with more confidence than you may feel.

1.6   Seeking after excellence

In many aspects of business today, just achieving an adequate performance is not enough. Competitive pressures have never been greater and this has a ready parallel with noncommercial-sector organizations. A university, for example, has just as many pressures arising from the financial side of its operation as a company does in seeking profitability; indeed, some of its activities may operate on a straight commercial basis. Similar things could be said about other kinds of organizations, from charities to government departments. As standards improve – in design, quality, service, whatever – then the broader market has to keep up and the effort needed to stay ahead increases. Excellence must be sought as standard to have any hope of competing.

All this puts people on the spot and certainly extends the challenge of many a job.

In customer service, when competition was less, a pleasant manner and reasonable efficiency shone out. Now, those dealing with customers, on the telephone, say, must offer product knowledge, advice and service linked to very specific standards. They may well be required to answer the telephone promptly, send sales material to arrive the following day, etc. – and still do so in a way that customers find spontaneous, courteous, informed and specific, while they operate complex computer equipment and see to the necessary documentation as they go. This is no easy task. Nor is that of managing a section working in this way. Such a manager may need skills of administration, of computer systems, of marketing and customer care, coupled with detailed knowledge of the product and customers. But, whatever else is needed, people-management skills certainly are – and also the communication skills that are an inherent part of them.

A *laissez-faire* approach, one primarily allowing staff to work out their own methods and respond to the inherent customer pressure as they think fit, may not maximize effectiveness; though it may superficially seem an easier way of working for the manager in the short term.

Achieving excellent performance takes some real working at. Of course, the effects can be worthwhile – as above in terms of customer satisfaction and thus future sales – but this effectively squeezes a larger management job into the same amount of time. This too is a problem, with people in many companies reporting that there is more and more to do in the time available, and sometimes with fewer people on their team than in the past.

There is no room for errors in communication to be allowed to reduce effectiveness in such circumstances; and every reason to use communication itself to enhance team effectiveness in any way it can. Good communication is a resource to be maximized.

1.7 Summary

So, communication directly affects performance. There is every reason to make the best of it. In a busy life many communication errors, inadequacies or omissions occur not as a result of lack of knowledge or understanding, but as a result of a lack of thought. Matters are skimped. Memos or reports are sent without being given the benefit of proper preparation. Meetings are run ad hoc, without clear objectives or agendas. And things are said on the spur of the moment that people live to regret. Much of the problem is time. Other pressures seem to intervene and rushing something else is seen as a – maybe unfortunate – necessity. Yet sorting out what occurs if things go wrong takes time too – sometimes more than can be saved by rushing.

Of course communicating well takes time. But there is no need for it to take much more than it would to communicate *less* well. What is more, any small amount of extra time so spent can be easily justified in terms of the impact it has on the individual, those around them and the organization as a whole.

Communicating effectively with people means understanding and thinking of both what makes communication work and how others view and respond to the whole process. It is to the latter that we turn in the next chapter. Meantime the key issues here are:

- never to underestimate the power of communications for good or ill;
- to realize that everyone has a personal responsibility for communicating well and that mistakes or excellence by the individual can have wide impact;
- to see communications as something needing study and care and to be actively worked at (it is all too easy to assume communications are straightforward, forget to engage the brain – and suffer the consequences).

Chapter 2
What makes for effective communication

Once in a while it is worth going back to basics. Think for a moment. Think again about how you would explain to someone how you tie a necktie (without demonstrating). How many times have you heard someone in your office say something like, 'But I thought you said ...' in the past week? What is the difference between saying something is 'quite nice' and is 'rather nice'? And would you find anything that warranted only either description the least bit interesting?

Make no mistake: communication can be difficult.

Have you every come out of a meeting or put the telephone down on someone and said to yourself, 'What's the *matter* with that idiot? They don't seem to understand *anything*'? And, if so, did it cross your mind afterwards that maybe the difficulty was that you were not explaining matters as well as you could? No? Should it have done? Why, when you get a wrong number, is it never engaged? Sorry, I digress – enough questions.

Make no mistake: the responsibility for making communication work lies primarily with the communicator. So, no surprise, the responsibility for getting through to people, for making things clear and ensuring understanding is, simply, yours. Consider two important rules.

- The first rule about communication is: *never assume any kind of communication is simple*. Most of the time we spend in our offices is taken up with communicating in one way or another. It is easy to take it for granted. Occasionally, we are not as precise as we might be, but never mind: we muddle through and no great harm is done. Except that occasionally it is. Some communication breakdowns become out-and-out derailments. Often, where there is much hanging on it, communications must be got exactly right and the penalties for not so doing range from minor disgruntlement to, at worst, major disruption to productivity, efficiency or quality of work.
- So the second rule, which is that *everyone needs to take responsibility for their own communication*, to tackle it in a sufficiently considered manner to make it work effectively, must make particular sense for all those whose job involves managing or liaising with people.

To set the scene for everything that follows we will now consider certain key influences on whether communication works effectively or not, and then how to get over the difficulties they present. This provides a practical basis for action for any kind of communication.

2.1   The difficulties of making communication effective

If there are difficulties, and there surely are, it is not because other people, or work colleagues or whoever, are especially perverse. Communication is, in fact, *inherently difficult*. Let us consider why.

2.1.1   Inherent problems

In communicating with people, what you do is essentially a part of the process of making an organization work. In so doing, your intentions are clear; it is necessary to make sure people:

- hear what you say, and thus listen;
- understand, and do so accurately;
- agree, certainly with most of it;
- take action in response (though the action may simply be to decide to take note).

Such action could be a whole range of things, be it agreeing to spend more time on something, attend a meeting or follow specific instructions.

Consider the areas above in turn:

2.1.1.1   First objective: to ensure people hear/listen (or read)

Here difficulties include the following.

- People cannot or will not concentrate for long periods of time, so this fact must be accommodated by the way we communicate. Long monologues are out; written communication should have plenty of breaks, headings and fresh starts (which is why the design of this book is as it is); and two-way conversation must be used to prevent people thinking they are pinned down and have to listen to something interminable.
- People pay less attention to those elements of a communication that appear to them unimportant, so creating the right emphasis, to ensure that key points are not missed, is a key responsibility of the communicator.

In other words you have to work at making sure you are heard – to earn a hearing.

2.1.1.2   Second objective: to ensure accurate understanding occurs

Difficulties here include the following.

- **People make assumptions based on their past experience**: So you must make sure you relate to just that. If you wrongly assume certain experience exists, your message will not make much sense. (Imagine trying to teach someone to drive if they had never sat in a car: 'Press your foot on the accelerator.' 'What's that?')
- **Other people's jargon is often not understood**: So think very carefully about the amount you use and with whom. Jargon is 'professional slang' and creates a useful shorthand among people in the know, for example, in one organization

or one industry, but dilutes a message if used inappropriately. For instance, used in a way that assumes a greater competence than actually exists it will hinder understanding (and remember that people do not like to sound stupid and may well be reluctant to say, 'I don't understand' – something that can apply whatever the reason for a lack of understanding).

- **Things heard but not seen are more easily misunderstood**: Thus, anything that can be shown may be useful; so, too, is a message that 'paints a picture' in words.
- **Assumptions are often drawn before a speaker finishes**: The listener is, in fact, saying to themselves, 'I'm sure I can see where this is going' – and their mind reduces its listening concentration, focusing instead on planning the person's own next comment. This, too, needs accommodating and, where a point is key, feedback can be sought to ensure that concentration has not faltered and the message really has got through.

2.1.1.3   Third objective: to prompt action

…And to do so despite the following.

- **It is difficult to change people's habits**: Recognizing this is the first step to achieving it; a stronger case may need making than would be the case if this were not true. It also means that care must be taken to link past and future: for example, not saying, 'That was wrong and this is better' but, rather, 'That was fine then, but this will be better in future' (and explaining how changed circumstances make this so). Any phraseology that casts doubt on someone's earlier decisions should be avoided wherever possible.
- **There may be fear of taking action**: 'Will it work?' 'What will people think?' 'What will my colleagues think?' 'What are the consequences of its not working out?' And this risk avoidance is a natural feeling; recognizing this and offering appropriate reassurance are vital.
- **Many people are simply reluctant to make prompt decisions**: They may need real help from you and it is a mistake to assume that laying out an irresistible case and just waiting for the commitment is all there is to it.

In addition, you need one more objective.

2.1.1.4   Fourth objective: to stimulate feedback

The difficulties here are these.

- Some (all?) people sometimes deliberately hide their reaction. Some flushing out and reading between the lines may be necessary.
- Appearances can be deceptive. For example, phrases such as 'trust me' are as often a warning sign as a comment to be welcomed – some care is necessary.

The net effect of all this is rather like trying to get a clear view through a fog. Communication goes to and fro, but between the parties involved lies a filter: not all of the message may get through, some may be blocked, some may be warped or let through only with pieces missing. In part, the remedy to all this is simply watchfulness. If you appreciate the difficulties, you can adjust your communications style a little to compensate, and achieve better understanding as a result.

The figure below illustrates this.

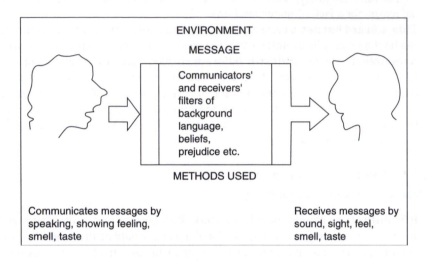

One moral is surely clear: communication is likely to be better for some planning. This may be only a few seconds' thought – the old premise of engaging the brain before the mouth (or writing arm) – through to making some notes before you draft a memo or report, or even sitting down with a colleague for a while to thrash through the best way to approach something.

We have already seen some possible antidotes to the inherent difficulties within the last few paragraphs, but are there any principles that run parallel and provide mechanisms to balance the difficulty and make matters easier? Luckily, the answer is that yes, there are.

2.2   Aids to effective communication

Good communication is, in part, a matter of attention to detail. Just using one word instead of another can make a slight difference. Actually, just using one word instead of another can make a *significant* difference (as you see!). And there are plenty of other factors that contribute, many of which are explored as this book continues. But there are also certain overall factors that are of major influence, and that can be used to condition your communications.

Four factors are key. We'll look at each in turn.

2.2.1   The 'What about me?' factor

Any message is more likely to be listened to and accepted if how it affects people is spelt out. Whatever the effect, in whatever way (and it may be 'ways'), people want to know, 'What's in it for me?' and 'How will it hurt me?' People are interested in both the potential positive and negative effects. Tell someone that you have a new computerized reporting system and they may well think the worst. Certainly, their reaction is unlikely to be simply, 'Good for you': it is more likely to be, 'Sounds like that'll be complicated' or 'Bet that'll have teething troubles or take up more time.' Tell them they are going to find it faster and easier to submit returns using the new system. Add that it is already drawing good reactions in another department, and you spell out the message and what the effects on them will be together, rather than leaving them wary or asking questions.

Whatever you say, bear in mind that people view it in this kind of way; build in the answers, and you avert their potential suspicion and make them more likely to want to take the message on board.

2.2.2   The 'That's logical' factor

The sequence and structure of communication are very important. If people know what it is, understand why it was chosen and believe it will work *for them*, then they will pay more attention. Conversely, if it is unclear or illogical, then they worry about it, and this takes their mind off listening. Something like this book provides an example: it might be possible to have a chapter investigating the fundamental principles of communication as Chapter 12, and a reason for it; but I doubt it. Certainly, readers would query it and look for a good reason.

Information is remembered and used in an order – you only have to try saying your own telephone number as quickly backwards as you do forwards to demonstrate this – so your selection of a sensible order for communication will make sense to people, and, again, they will warm to the message. Using an appropriate sequence helps gain understanding and makes it easier for people to retain and use information; as with much of what is said here, this is especially true for a technically oriented message.

Telling people about this is called *signposting*: flagging in advance either the content or nature of what is coming next; one important form of this is describing a brief agenda for what follows.

Signposting is a very useful device. Say, 'Let me give you some details about what the reorganization is, when the changes will come into effect and how we will gain from it' – and, provided that makes sense to your listener, they will want to hear what comes next. So tell them about the reorganization and then move on. It is almost impossible to overuse signposting. It can lead into a message, giving an overview, and also separately lead into subsections of that message. Sometimes it can be strengthened by explaining why the order has been chosen: 'Let's go through it chronologically – perhaps I could spell out …'

Whatever you have to say, think about what you say first, second, third and so on and make the order you choose an appropriate sequence for whomever you are communicating with.

2.2.3   The 'I can relate to that' factor

Imagine a description: 'It was a wonderful sunset.' What does it make you think of? Well, a sunset, you may say. But how do you do this? You recall sunsets you have seen in the past and what you imagine draws on that memory, conjuring up what is probably a composite based on many memories. Because it is reasonable to assume that you have seen a sunset, and enjoyed the experience in the past, I can be fairly certain that a brief description will put what I want into your mind.

It is, in fact, almost impossible not to allow related things to come into your mind as you take in a message. (Try it now – and *do not* think about a long, cool refreshing drink. See?) This fact about the way the human mind works must be allowed for and used to promote clear understanding.

On the other hand, if you were asked to call to mind, say, the house in which I live and yet I describe it to you not at all, then this is impossible – at least unless you have been there or discussed the matter with me previously. All you can do is guess, wildly perhaps, that, 'All authors live in garrets' or 'All authors are rich and live in mansions' (wrong on both counts!).

So, with this factor also inherent to communication, it is useful to try to judge carefully people's prior experience; or, indeed, to ask about it if you have not known them for long and you are unsure of their past experience. You may also refer to it with phrases linking what you are saying to the experience of the other person – for example, saying things such as 'this is like …', 'you will remember …', 'do you know so and so?', 'this is similar, but …' – all designed to help the listener grasp what you are saying more easily and more accurately.

Beware of getting at cross purposes because you think someone has a frame of reference for something that they do not; link to *their* experience and use it to reinforce your message.

2.2.4   The 'Again and again' factor

Repetition is a fundamental help to grasping a point. Repetition is a fundamental help to …Sorry. It is true, but it does not imply just saying the same thing, in the same words, repeatedly. Repetition takes a number of forms:

- things *repeated in different ways* (or at different stages of the same conversation);
- points made in *more than one manner*: for example, being spoken and written down;
- using *summaries or checklists* to recap key points;
- reminders *over a period of time* (maybe varying the method, phone, email or meeting).

This can be overdone (perhaps as in the introduction to this point above), but it is also a genuinely valuable aid to getting the message across, especially when used with the other factors now mentioned. People really are more likely to retain what they take in more than once. Enough repetition.

2.3   Positioning your communication

So far in this chapter the principles outlined have been general; they can be useful in any communication. But exactly whom you communicate with is important. Consider staff, reporting to a manager, as a special category. If you want people to work willingly, happily and efficiently for you, one useful approach to any staff communication is to remember not to allow your communication style to become too introspective – if you want to influence them, relate to them in a way that makes *them* the important ones. Although you speak *for* the organization, staff members do not appreciate an unrelieved catalogue that focuses predominantly on your side of things:

- 'the organization is ...'
- 'we have to make sure ...'
- 'I will be able to ...'
- 'our service in the technical field is ...'
- 'my colleagues in research ...'
- 'our organization has ...'

...and so on. Any such phrases can be turned round to focus on the people thus: 'You will find this change gives you ...'; 'You will receive ...'; 'You can expect that ...'

A slight mixture is, of course, necessary, but a predominantly introspective approach always seems somewhat relentless. And it is more difficult when phrasing things that way round for you to give a real sense of tailoring what you say to the individual: introspective statements sound very general. Using the words *you* and *yours* (and similar) at the start of a message usually works well, and once this start is made it is difficult for you to make what you say sound introspective.

2.4   Projecting the right impression

Having made a point about not sounding too introspective, I have to say that, on the other hand, you do need to be concerned about the image you put across, because there is a good deal more to it than simply sounding or appearing pleasant.

Some factors are largely common. You will probably want to include a need to appear:

- efficient;
- approachable;
- knowledgeable (in whatever ways circumstances dictate);
- well organized;
- reliable;
- consistent;
- interested in your staff;
- confident;
- expert (and able to offer sound advice).

For example, people like to feel they are working for someone competent, someone they can respect. Fair enough. But the thing to note is that there is a fair-sized list of characteristics that are worth getting over, and all of them are elements that can be *actively* added to the mix, as it were. You can *intend* to project an image of, say, confidence and make it more than you feel, or of fairness when you want it to be absolutely clear that this is what you are being. Projecting the right mix – and balance – of characteristics to create the right image is important. There is some complexity involved here and thus it is another aspect of the whole process that deserves some active consideration. Anyone, whatever their role, can usefully think through the most suitable profile for them in this way.

In addition, you must often have a clear vision of the kind of way you want to project the organization you represent and the department or function you are in, and project that too. This is especially important when you are dealing with people with whom you have less than day-to-day contact, those in other departments, for instance.

Consider whether you should put over an appearance of:

- innovation;
- long experience and substance;
- technical competence;
- having a very human face;
- confidence.

Again, you must decide the list that suits you, and emphasize your intended characteristics as appropriate to create the total picture that is right for whomever you communicate with. This is often no more than just a slight exaggeration of a characteristic, but can still be important.

In all these cases, different levels and types of person will need different points emphasizing in different ways. For example, some people may warm to an experienced manager with apparent concern for their staff. If so, then any qualities creating that impression can usefully be stressed. Others may seek more weight, so a style with more telling involved makes sense for them, and you will need to project appropriate clout to make it stick.

Individually, all the factors mentioned in this chapter are straightforward. Any complexity in making communication work comes from the need to concentrate on many things at once. Here, habit can quickly come to our assistance. There is a danger in this, too, however. Unless you maintain a conscious overview, it is easy to slip into bad habits or, by being unthinking – and making no decision rather than making the wrong decision – allow the fine tuning that makes for good communication to go by default. Remember, just a word or two can make a difference. A complete message delivered in an inadequate manner may cause chaos.

Two other key factors have not yet been given sufficient weight. As we communicate we have to work at what occurs in both directions.

2.5   Amplifying communications

One or two added factors are worth consideration. Clearly, listening is vital, but it is not enough just to say that. You need to make listening another *active* process. This involves care and concentration.

Listening has to make a difference to a conversation. Above all, adapting how you proceed in the light of the information other people give is key, and an important element of this is *being perceived to do so*. Few things endear you to other people quite so much as being a good listener. It is a factor that needs only a little thought and can quickly become a habit, one that is important throughout the piece as we move on and we return to it in the next chapter.

Before moving on let's consider some comment on a technique that is common in providing assistance with many of the forms of communication we will go on to review in succeeding chapters: that of questioning techniques.

Many communications situations need to be clarified by the asking of questions. Unless you know the facts, unless you know what people think and, most important of all, unless you know *why* things are as they are, taking the process on may be difficult or impossible. How do you resolve a dispute if you do not really understand why people are at loggerheads? How do you persuade people to action when you do not know how they view the area in which you want them to get involved? How do you motivate if you do not know what is important to people or what worries them? The answer in every such case might be stated as 'with difficulty'. Questions create involvement, they get people talking and the answers they prompt provide the foundation for much of what makes communication successful.

But questioning is more than just blurting out the first thing that comes to mind: 'Why do you say that?' Even a simple phrase may carry overtones and people wonder if you are suggesting they should *not* have said that, or if you see no relevance for the point made. In addition, many questions can easily be ambiguous. It is all too easy to ask something that, only because it is loosely phrased, prompts an unintended response. Ask, 'How long will that take?' and the reply may simply be 'Not long.' Ask, 'Will you finish that before I have to go to the meeting at eleven o'clock?' and, if your purpose was to be able to prepare for the meeting accordingly, then you are much more likely to be able to decide exactly what to do.

Beyond simple clarity, you need to consider and use three distinctly different kinds of question.

1. **Closed questions**: These prompt rapid yes or no answers, and are useful both as a starting point (they can also be made easy to answer to help ease someone into the questioning process) and to gain rapid confirmation of something. Too many closed questions on the other hand create a virtual monologue in which the questioner seems to be doing most of the talking, and this can be annoying or unsatisfying to the other person

2. **Open questions**: These are phrased so that they cannot be answered with a simple yes or no and typically begin with words such as *what*, *where*, *why*, *how*, *who* and *when* and phrases such as 'Tell me about ...' Such questions get people talking, they involve them and they like the feel they give to a conversation. Prompting a fuller answer and encouraging people to explain means they also produce far more information than from closed questions

3. **Probing questions**: These are a series of linked questions designed to pursue a point: thus, a second question that says, 'What else is important about ...?' or a

phrase like 'Tell me more about ...' gets people to fill out a picture and can thus produce both more detail and the *why* that lies beyond more superficial answers.

Many a communication is made to succeed by the simple prerequisite of starting it with some questions. It is important to give sufficient time to the questioning process when finding out is necessary. It may also be important to give the clear impression to other people that sufficient time is being given to something. This may indicate, say, the importance with which something is regarded; and the reverse may give the wrong impression – say of lack of concern. Both may be important. This is something that it may sometimes be useful to spell out with, say, 'I want to go through this thoroughly. I can take an hour or so now and if that proves inadequate we can come back to it. Let's see how we get on.'

2.6   Summary

Two approaches are essential to everything in business communication:

1. that you recognize the inherent problems that exist and make communication less certain (and do not assume the process is straightforward);
2. that you recognize also that your communication needs actively to aim to get over (or at least minimize) these problems and act accordingly.

Unless this is accommodated, adding the further techniques involved in any particular form of communication is going to allow effectiveness to be diluted.

Chapter 3
Prerequisites for success: preparation and listening

3.1 A fundamental truth

Enough has been said about the difficulties of communicating effectively to demonstrate that anything and everything that helps it go well is worth considering. Here we go further and consider elements that are absolute fundamentals to getting it right. Much of what is said here is generic: it helps any communication. Beyond that, there is a necessity to adopt specific approaches to specific tasks; so writing a report, for instance, has some unique elements to it.

Note: This chapter is thus linked strongly to others and it may be worthwhile referring back to it as you read other parts of this book; this is suggested in a variety of specific places throughout the book.

3.2 Listening

This may be an obvious one, but it stands some real consideration.

Do not look back, but can you remember what particular kind of communication was mentioned as an example in the first paragraph of this chapter? Or can you describe either of the two summary points at the end of the last chapter? If not (and be honest), then consider that the principle is similar – without concentration we do not take in every detail of what we are reading, or hearing.

The key thing, then, is to regard listening as an *active* process. It is something we all need to work at. What does this mean? There are what are perhaps a surprising number of ways in which we can focus and improve our listening – and the retention of information, including details crucial to understanding, that it enables. These include the need to do the following.

- **Want to listen**: This is easy once you realize how useful it is to the communication process.
- **Look like a good listener**: People will appreciate it and, if they see they have your attention and feedback, will be more forthcoming.
- **Understand**: It is not just the words but the meaning that lies behind them that you must note.

- **React**: Let people see that you have heard, understood and are interested. Nods, small gestures and signs and comments will encourage the other person's confidence and participation – right?
- **Stop talking**: Other than small acknowledgements, you cannot talk and listen at the same time. Do not interrupt.
- **Use empathy**: Put yourself in the other person's shoes and make sure you really appreciate their point of view.
- **Check**: If necessary, ask questions promptly to clarify matters as the conversation proceeds. An understanding based even partly on guesses or assumptions is dangerous. But ask questions diplomatically; avoid saying, 'You didn't explain that properly.'
- **Remain unemotional**: Too much thinking ahead – 'However can I overcome that point?' – can distract you.
- **Concentrate**: Allow nothing to distract you.
- **Look at the other person**: Nothing is read more rapidly as uninterest than an inadequate focus of attention – good eye contact is essential (in negotiating, for instance, lack of it will always be read as deviousness).
- **Note particularly key points**: Edit what you hear so that you can better retain key points manageably.
- **Avoid personalities**: Do not let your view of someone as a person distract you from the message, or from dealing with them if that is necessary.
- **Do not lose yourself in subsequent arguments**: Some thinking ahead may be useful; too much and you suddenly may find you have missed something.
- **Avoid negatives**: To begin with clear signs of disagreement (even a dismissive look) can make the other person clam up and destroy the dialogue.
- **Make notes**: Do not trust your memory, and, if it is polite to do so, ask permission before writing their comments down.

Make no mistake, if you listen – *really* listen – then everything that follows will be a little easier and more certain.

3.3   Preparation: a moment's thought

We should not think that having to prepare implies some sort of weakness. For instance, the 'born' public speaker, effortlessly sailing through a presentation, is probably able to give this impression only because they are well prepared. It needs doing; the job is to make sure it is well done and is also done productively – good preparation should save time overall.

Whether you are to write a report, make a presentation or undertake something simpler, such as writing an email or a letter, the process is essentially similar. What does change is the complexity and the time that preparation takes. In order to describe an approach and make it seem real, the following relates primarily to making a presentation: often something written links to something presented and, whatever is done, *the key approaches always apply*.

3.4　Setting objectives

Whatever you may need to communicate and however it is to be done, its purpose must be clear. You must be able to answer the question, 'Why am doing this?' And set out a purpose, one that always needs to involve you and the recipients of your message and describes what effect you aim to have on them. Remember that communication can have many overall intentions (to inform, motivate and more – described in Chapter 1), and that these are not mutually exclusive. The more different intentions there are, the more preparation must ensure all will be fulfilled.

Objectives need not only to be clear, but spelt out in sufficient detail (certainly in your own mind and sometimes for others). They must act as a genuine guide to what you will do. They also need to reflect not just what you want, but the audience's view also.

Often a much-quoted acronym can provide a good guide here: SMART. This stands for:

- Specific
- Measurable
- Achievable
- Realistic
- Timed

As an example, you might regard objectives linked to your reading a latter chapter – on presentations – as to achieve the following.

- Enable you to ensure your presentations come over in future in a way that audiences will see as appropriate and informative (*specific*).
- Ensure (*measurable*) action takes place afterwards (here you might link to any appropriate measure: from agreements or actions that group members take or commit to the volume of applause received!).
- Be right for you: sufficient, understandable information in manageable form that really allows you to change and improve what you do later (an *achievable* result).
- Be *realistic*, that is desirable – hence a short text (if it took you several days to read the effort might prove greater than any benefit coming from doing so).
- Provide *timing* – always a good factor to include in any objective. When are you going to finish reading this chapter? When is your next presentation? How far ahead of it should you prepare?

So, ask yourself whether you are clear in this respect before you even begin to prepare. If you know *why* the presentation must be made, and *what* you intend to *achieve* then you are well on the way to success. Time spent sorting this, and making sure you have a clear vision of what the objectives are, is time well spent. It may take only a few moments, but is still worth doing. Or it may need more thought and take more time. So be it. It is still worth doing and, in any case, may well save time on later stages of preparation.

With your purpose clear, and a constant eye, as it were, on the audience, you can begin to assemble your message.

3.5   Deciding the message

There is more to this than simply banging down the points in sequence, something that was hinted at early in this chapter. A more systematic approach is necessary; indeed, a more systematic approach can quickly become a habit of preparing in a way that promptly and certainly enables you to deliver what you want.

The following provides a full description of a tried and tested approach. This describes the fullest degree of preparation necessary, but it is important to stress that this is not offered as something that must be followed slavishly. The important thing is to find, experiment with and refine and then use a method that suits *you*. In addition, practice and experience, or other factors such as familiarity with your chosen topic, may well allow you to adopt a 'shorthand' version of these approaches, which is quicker, but still does for you the total job that is necessary at this stage.

There is a need here to take one point at a time (if only because there is no other way to proceed).

First, a six-stage approach majors on helping sort out *what the message is to be*, *what you* need to say in our presentational example (and what you should not say also). Here also we investigate more about *how* you will put the message across. Both link to the structure involved: what comes first, second and third and what constitutes the beginning, the middle and the ending.

There is something of the chicken and egg here. Does preparation or structure logically come first? Both are important, both are interrelated. The sequence chosen here works well and is intended to show you how to put a presentation together as it would need to be done in a real-life situation. The details and the sequence can equally apply to something such as writing a report or proposal, and in less elaborate form to much else besides. So, on to the detail of assembling the message.

3.6   Putting it together

It is not necessary only to 'engage the brain before the mouth', but also vital to think through, in advance, what a presentation must contain – and not contain, for that matter. The following process of thinking through and preparation is recommended solely by its practicality and can be adapted to cope with any sort of presentation, of any length or complexity and of any purpose.

Many communications fail or their effectiveness is diluted because preparation is skimped. Accepting that preparation takes time and building this into the business of the workplace is the first step to being a good communicator. In the long run, it saves time, in part on the old premise that, while there is never time to do things properly, there always has to be time made available to sort out any mess caused by their inadequacies.

There are six stages (described, in part, by continuing to use the presentations example). The very best way of linking the principles described here to real life is to go through them with some personal project, such as a presentation you must make, in mind and link this to the approach that follows.

3.6.1   Stage 1: Listing

Forget about everything such as sequence, structure and arrangement, and just concentrate on and list – in short note (or keyword) form – every significant point that the presentation might usefully contain. Give yourself plenty of space (something larger than the standard A4 sheet is often useful: it lets you see everything at one glance). Set down the points as they occur to you, almost at random across the page. For something simple this might result only in a dozen words, or it might be far more.

You will find that this is a good thought prompter. It enables you to fill out the picture as one thought leads to another, with the freestyle approach removing the need to pause and try to link points or worry about sequence. With this done – and with some messages it may only take a short time – you have a full picture of the possibilities for the message in front of you and you can move on to the second stage.

3.6.2   Stage 2: Sorting

Now, you can review what you have noted down and begin to bring some order to it, deciding:

- what comes first, second and so on;
- what logically links together, and how;
- what provides evidence, example or illustration to the points.

At the same time, you can – and probably will – add some additional things and have second thoughts about other items, which you will delete, as well as amending the wording a little if necessary You need to bear in mind here what kind of duration (or length) is indicated, and what will be acceptable.

This stage can often be completed in a short time by simply annotating and amending the first-stage document. Using a second colour makes this quick and easy, as do link lines, arrows and other enhancements of the original notes.

At the same time, you can begin to catch any more detailed element that comes to mind as you go through (including ways of presenting as well as content), noting what it is at more length on the page or alongside.

3.6.3   Stage 3: Arranging

Sometimes, at the end of Stage 2, you have a note that is sufficiently clear and from which you can work directly in finalizing matters. If it can benefit from clarification, however, it may be worth rewriting it as a neat list; or this could be the stage where you type it and put it on screen if you are working that way and want to be able to print something out in due course.

Final revision is possible as you do this. Certainly you should be left with a list reflecting the content, emphasis, level of detail and so on that you feel is appropriate. You may well find you are pruning a bit to make things more manageable at this stage, rather than searching for more contents and additional points to make.

3.6.4   Stage 4: Reviewing

This may be unnecessary. Sufficient thought may have been brought to bear through earlier stages. However, for something particularly complex or important (or both) it may be worth running a final rule over what you now have down. Sleep on it first perhaps – certainly avoid finalizing matters for a moment if you have got too close to it. It is easy to find you cannot see the wood for the trees.

Make any final amendments to the list (if this is on screen it is a simple matter) and use this as your final 'route map' as preparation continues.

3.6.5   Stage 5: Prepare the 'message'

In our example this would be speakers' notes (see more about these in the appropriate chapter). If the job were to write something, then this is where you actually write it. Now you can turn your firm intentions about content into something representing not only *what* will be said, but also *how* you will put it over. One of the virtues of the procedure advocated here is that it stops you trying to think about what to say and how to say it at the same time; it is much easier to take them in turn. This fifth stage must be done carefully, though the earlier work will have helped to make it easier and quicker to get the necessary detail down.

Here are a couple of tips.

- If possible, *choose the right moment*. There seem to be times when words flow more easily than others (and it may help literally to talk it through to yourself as you go through this stage). Certainly, interruptions can disrupt the flow and make the process take much longer, as you recap and restart again and again. The right time, uninterrupted time in a comfortable environment – all help.
- *Keep going*. By this I mean do not pause and agonize over a phrase, a heading or some other detail. You can always come back to that; indeed, it may be easier to complete later. If you keep going you maintain the flow, allowing consistent thinking to carry you through the structure to the end so that you can 'see' the overall shape of it. Once you have the main detail down, then you can go back and fine-tune, adding any final thoughts to complete the picture. The precise format of notes can be very helpful, something that is investigated later.

3.6.6   Stage 6: A final check

A final look (perhaps after a break) is always valuable. This is also the time to consider rehearsal: either talking it through to yourself, a tape recorder or a friend or colleague; or going through a full-scale 'dress rehearsal'.

Thereafter, depending on the nature of the presentation, it may be useful – or necessary – to spend more time, either in revision or just reading over what you plan to do. You should not overdo revision at this stage, however; there comes a time simply to be content that you have it right and stick with it. If preparing a written document, it is here that any necessary editing takes place. And there will be some editing – few, if any people, write without the need to fine-tune the text to produce a final version.

This whole preparation process is important and not to be skimped. Preparation does get easier, however. You will find that, with practice, you begin to produce material that needs less amendment and that both getting it down and the subsequent revision begin to take less time.

At the end of the day, as has been said, you need to find your own version of the procedures set out here. A systematic approach helps, but the intention is not to overengineer the process. What matters is that you are comfortable with your chosen approach – that it works for you. If this is the case, then, provided it remains consciously designed to achieve what is necessary, it will become a habit. It will need less thinking about, yet still act to guarantee that you turn out something that you are content with and meets the needs – whatever they may be.

3.7   Summary

Preparation is a vital part of communicating. At its simplest it is merely a moment's constructive thought. More often more is necessary. The key issues are:

- always to precede any thinking by devising a clear objective;
- to prepare messages with a clear idea of what intentions they reflect (informing, persuading, etc.);
- to think matters through systematically and separate deciding *what* you will say (or put over in whatever way), from *how* you will put things and thus the precise language you will use;
- to give this process sufficient time and, if possible, build in some pauses so that you do not become unable to see the wood for the trees;
- to be prepared to fine-tune the message to get it right.

Chapter 4
Being persuasive: getting agreement from others

Communication, as there is a danger of overstating, is not easy. Difficulties abound when we are simply trying to impart information and do it clearly; it is even more difficult if you intend to persuade someone to do something that you want, and that perhaps they do not, at least at first.

Although it was said earlier that there are no magic formulae, it is not exaggerating to suggest that preparation comes close. Certainly, the person who runs rings around others as it were is probably not inherently persuasive – more likely they understand how this sort of communication works and 'do their homework'.

Remember that preparation does a number of things. Not least, the thinking involved makes executing communication easier – it provides a 'route map' to guide you through what needs to be done.

While avoiding undue repetition, and coupling that to recommending that you check carefully the full implications of preparing in the previous chapter, it is worth highlighting the four key stages that must be gone through:

1. *setting objectives* (SMART objectives);
2. *checking the facts* (ensuring you have the facts to hand, whether about the topic of your planned persuasion, the people you intend to persuade or whatever else);
3. *planning the meeting* (so that you are able to go through things as closely as possible to what you consider the ideal way, without making other people feel they are being led dictatorially);
4. *backing up what you will say* (assembling the elements of the case you will make).

With this thinking done, you are ready to communicate. You know:

*   precisely what you are aiming at;
*   how you intend to go about presenting your case;
*   something about the other person – and therefore their likely reactions;
*   what you will use to exemplify your case;
*   what problems may occur and, broadly, how you will deal with them.

4.1 The logistics

It may also be important to think about certain other matters. For example, how long are you likely to have? It is no good planning a blindingly convincing case that takes half an hour to deliver, if you realistically will have only half that time. Or where will you be? Will there be room for you to lay all the materials you plan to use out on the table? Note: if you have to deal other than face to face, remember that other techniques may be involved, for example writing or 'on-your-feet' presentation, and that this can add an additional dimension to the process that also needs some thought.

You can never know, of course, exactly how things will go and your planning must not act as a straitjacket, but allow you to retain an inherent flexibility. But having all this clear in your mind will certainly help; what is more, it adds another important element to the equation – and to your chances of getting your own way: confidence. If you are clear in your own mind of the path ahead, and have to make less of it up as you go along, then what you will do will be easier – and more certain.

4.2   Do unto others

Preparation may not be a magic formula that guarantees success, but it will certainly help. So too does the approach described next. The key premise is absurdly simple.

You are more likely to be persuasive if you approach the process as something that relates as much to what *other people* want and how they think as to what you want to do yourself. Indeed, this perspective must underlie everything you do, so the starting point is to think through how being persuaded looks from the other person's point of view.

There are three aspects to other people's views that need to be considered and then borne in mind: (1) how they feel, (2) what they want and (3) how they go about making a decision to agree to act, or not.

4.2.1   Others' feelings

Most often, people recognize very quickly when they are in a situation where someone is trying to persuade them of something. Their instinctive reaction may be to dislike the idea of it: 'I'm not being made to do anything.' However, once they begin to appreciate what is being asked of them, their feelings may be positive or negative, or indeed a mixture of both. Positive reactions are clearly easier to deal with, and can work for us.

In an obvious case – persuading someone to do something they will clearly find beneficial – they may start to see it as a good idea almost at once. So, a manager saying to someone that they want to discuss some changes to their work portfolio that will make their life easier and put them in line for a salary increase will likely find them all ears. This does not mean that they will not be on the lookout in case what is being suggested is not 100 per cent good, but essentially their thinking will tend to be positive. In this case, there may well be no difficult implications for the person doing the persuading, other than to aim to build on the goodwill that is already starting to exist.

But the opposite may, perhaps more often, be the case. A variety of negative feelings may arise, immediately or as you get into making your case, and if so then you need to be sensitive to what is happening and seek to position what you do in light of it.

The following sets out some examples of how people might feel and what they might think.

- **Insecure**: This sounds complicated, I am not sure I will know how to decide or what view to take.
- **Threatened**: Things are being taken out of my hands. *I* should decide this, not be pushed into something by someone else.
- **Out of control**: If I make the wrong decision I may be in trouble. Any decision involves taking a risk and things could backfire on me.
- **Worried**: You're suggesting changes – does that imply I was at fault before? I don't like that implication.
- **Exposed**: This discussion is getting awkward. I'm being asked to reveal facts or feelings that I would rather not discuss.
- **Ignorant**: You're using your greater knowledge to put me on the spot. I don't feel confident in arguing the point, though I am not convinced.
- **Confused**: I ought to understand, but you're not making things clear – or letting me get any clarification.
- **Sceptical**: You make it sound good but, then, it's what *you* want. Maybe the case is not as strong as it seems.
- **Misunderstood**: I don't believe the case you make takes my point of view into account – it's all right for you but not for me.
- **Suspicious**: – People with 'something to sell' always exaggerate and are interested only in what they want – I'm not going to be caught out by this.

A moment's thought quickly suggests all such feelings are understandable, but if they are overlooked – if you go ahead as if your message should be received with open arms when in fact such reactions exist – you will hit problems. If thoughts such as this *are* in people's minds, then they act to cloud the issue and may make it more difficult to see the logic of something you are suggesting. It is not enough to be clear or to present what seems to you an obviously strong case – the other person must see it as something with which they can willingly go along.

4.2.2   What others want

What people want may vary enormously, of course. It will relate back to their situation, views, experience and prejudices. It may reflect deep-seated, long-held views or be more topical and transient – or both. Sometimes you know in advance what people want. On other occasions it comes out in the course of conversation, or you need to ferret it out as you go along. It can be complicated – with a number of different 'wants' involved together (some of which may be contradictory) – and thus needs some thought to keep it in mind. But understanding and responding to people's desires are an important part of being persuasive.

Panel 4.1   Example

A simple example will make what is involved here clearer (this will be referred
to throughout this chapter). Imagine you have to make some sort of formal
presentation jointly with a colleague at work. You want to persuade them to set
aside sufficient time, in advance, to rehearse the presentation together to make
sure it goes well. What might they want? Maybe the following.

1. **Make sure it goes well**: As you do, but maybe they are more confident of
   making it go well than you are.
2. **Minimize time spent in preparation**: Like you again, no doubt, but perhaps this blinds them to the need for rehearsal, which they might see as a
   sledgehammer to crack a nut.
3. **Leave preparation to the last minute**: Maybe because other tasks have
   greater short-term urgency, or seem to have.
4. **Outshine you on the day**: They might be more intent on scoring personal
   points with someone, than on making the overall event go well.

These are examples only, many feelings might be involved depending on
the nature of the presentation, how important it is and how they feel about it.
One thing is clear however: such wants make a difference to the likelihood
of your getting agreement. Even in a simple example like this the individual
viewpoints are clear: (1) you both want it to go well, but take differing views of
what is necessary to make this happen; (2) in general you want the same thing,
but would define the amount of time that constitutes the minimum differently;
(3) here you differ; and (4) there are very personal wants that are, to a degree,
outside of the main objective involved: that of making your presentations work
seamlessly together.

There is a need to balance the differing viewpoints if agreement is to be
forthcoming. If you are the persuader, you feel your viewpoint is right – or at
least the most appropriate option. How do you move them towards it? Clearly,
doing so involves their adjusting their intentions. You do not have to persuade
them to change their views completely. For instance, they may always see it as
easier to do whatever preparation is involved at the last minute, but may still
agree to set a time when you want or – compromise may often be involved –
somewhere between your two views.

4.2.3   How decisions are made

A good definition of selling is the simple statement that selling is *helping people to
buy*. Similarly, whatever the commitment is you are looking to secure, any persuasive
process of obtaining it is best viewed as one that *assists people to make a decision*,
and that, at the same time, encourages them to make it in favour of whatever option
you are suggesting. In a purchasing situation the choices involve competition: if you

are buying a washing machine, say, then you may find yourself having to decide whether to purchase the Hoover, the Indesit, the Bosch or whatever (as well as decide where to buy it from and what to pay). In other situations, choice is always involved.

In the presentation example used earlier, choice is a key element. Your imagined colleague will decide between rehearsing and not rehearsing; between rehearsing earlier and rehearsing later; between doing so in a way that helps them and doing so in a way that helps both of you – and so on. Doing nothing may seem, in many circumstances, an attractive option and needs as much arguing against as any other.

It follows that, if a process of decision-making is inherently involved, you should not fight against it. The intention should be to *help* it. Persuasive communication is not something you *direct at other people*. It is something you *engage in with them*. The difference is crucial, and anything that leads you to see it as a one-way process is likely to end up making the task you seek to accomplish more difficult. So far, so good, but how exactly do people make decisions?

The answer can be summed up succinctly, people:

- consider the *options*;
- consider the *advantages and disadvantages* of each;
- *weigh up* the overall way in which they compare;
- select what seems to be, on balance, the *best course of action* to take.

This does not mean finding and selecting an option with no downsides; realistically this may simply not be possible. It means assessing things and selecting an acceptable option, one where the pluses outweigh the minuses. The analogy of a balance or weighing scales is a good one to keep in mind. Imagine a set of old-fashioned scales with a container on each side. One contains a variety of plus signs, the other minuses. The signs are of different sizes because some elements of the argument are more important than others – they weigh more heavily on the scales. Additionally, some signs represent tangible matters. Others are more subjective, just as, in the presentation example above, achieving the right result from it (say getting agreement to a 10 per cent increase on a budget) is tangible. But an individual's desire to increase their status within an organization through the way they are perceived as a presenter is *intangible*. Intangible some points may be, but they can still be a powerful component of any case.

A final point completes the picture here: some decisions are more important than others and therefore may be seen to warrant more thought. Where a decision is of this sort, people may actively want it to be *considered*. They want to feel that the process of making it has been sensible and thorough (and therefore their decision is more likely to be a good one); and they may want other people (their manager, say) to feel the same. In either case, this feeling may lengthen the process of persuading them.

4.3   The thinking involved

This weighing-scales analogy is worth keeping in mind. It can act as a practical tool, helping you envisage what is going on during what is intended to be a persuasive conversation. Beyond that, it helps structure the process if you also have a clear idea of the sequence of thinking involved in this weighing-up process.

Psychologists have studied the process. One such way of looking at it is to think of people moving through several stages, as it were saying the following to themselves.

- **I matter most.** Whatever you want me to do, I expect you to worry about how I feel about it, respect me and consider my needs.
- **What are the merits and implications of the case you make?** Tell me what you suggest and why it makes sense (the pluses) and whether it has any snags (the minuses) so that I can weigh it up; bearing in mind that few, if any, propositions are perfect.
- **How will it work?** Here they additionally want to assess the details not so much about the proposition but about the areas associated with it. For instance, in the presentation example: 'Does rehearsal mean I have to give you a written note of what I will say?' – a time-consuming chore and thus seen as negative. Conversely, if this is unnecessary, then that fact may go on the plus side of the balance.
- **What do I do?** In other words what action – exactly – is now necessary? This too forms part of the balance. If something early on in this book persuaded you that it might help you, you may have bought it. In doing so you recognized (and accepted) that you would have to read it and that this would take a little time. The action – reading – is inherent in the proposition and, if you were not prepared to take it on, might have changed your decision.

It is after this thinking is complete that people will feel they have sufficient evidence on which to make a decision. They have the balance in mind, and they can compare it with that of any other options (and, remember, some choices are close run with one option only just coming out ahead of others). Then they can decide, and feel they have made a sensible decision on a considered basis.

This thinking process is largely universal. It may happen very quickly and be almost instantaneous – the snap judgement. Or it may take longer, and that may sometimes indicate days or weeks (or longer!) rather than minutes or hours. But it is always in evidence. So there is always merit in setting out your case in a way that sits comfortably alongside it. Hence *helping the decision-making process*.

Before we move on to how to orchestrate the actual communication and make it persuasive, there is one other factor that also needs to be deployed in a way that respects the other person.

4.4   Your manner

Your communication style no doubt reflects your personality. Certainly there is no intention here to suggest that you forget or disguise that and adopt some contrived manner in the belief that this will make you more persuasive; it will not.

On the other hand you do need to think about how you come over. Will it help your case to be seen to be: knowledgeable, expert, caring, friendly, responsive, adaptable, secure, well organized, efficient, forward-thinking, confident, interested (particularly in the other person or the topic of discussion), respectful, consistent, reliable or whatever? (And what do you *not* want to appear?) Is it important that you display an attention to detail, a respect for the other person's time or that you 'look the part' in some way? Many factors might be involved and such a list could doubtless be extended.

The point is not only that there are many such factors that can be listed, but also that they are *all options*. You can *elect* to come over as, say, confident or expert (to some degree even if you are not!). You can emphasize factors that are important to the other person; indeed you need to anticipate what these will be. If they want to dot every *i* and cross every *t*, so be it; you need to become the sort of person who does just that if it will allow you to get your own way in the end.

This is not so contrived, just an exaggerated version of what we do all the time as we communicate with different kinds of people – for example at opposite ends of the organizational hierarchy. Again, a little thought ahead of actually communicating can allow you to pitch things in the right kind of way, so that your manner enhances the chances of getting your own way rather than negating them.

Two factors are especially important here.

1. **Projection**: This word is used to encapsulate your approach, personality, authority, clout and the whole way in which you come over.
2. **Empathy**: This is the ability to see things from other people's points of view. More than that, it is the ability to *be seen* to see things from other people's points of view.

These act together. Too much projection and you come over as dictatorial and aggressive. Too little empathy and you seem insensitive and uncaring. You need to deploy both, and they go well together. Sufficient empathy softens what might otherwise be seen as a too powerful approach, and makes the net effect acceptable. This may necessitate only a few words being changed, with an unacceptable 'I think you should do this' being replaced by something like 'Given that you feel timing is so important, you may want to do this.'

At this point, well prepared and with a close eye on how the other person will consider your suggestion, and in what way they will go about coming to a decision to go along with it or not, we can turn to how to structure and put over a persuasive case.

4.5 Making a persuasive case

Your communication may take various forms, but let us consider the ubiquitous meeting. This may be formal, with two people (or more) sitting comfortably around a desk, or happen on the move (walking from the office to the pub for lunch) and sometimes it will occur in more difficult circumstances (a discussion, on your feet, in a factory with noisy machinery clattering in the background). In every case the objective is the same: to create a considered message that acts persuasively to prompt someone to take whatever action you seek.

In order to be able to proceed on a considered basis, you need to draw on a clear view of what is happening during such a meeting. We will dissect the process to tease out the key issues.

4.5.1   First impressions last

The manner you adopt and the preparation you have done will both contribute to your making a good start. So too will your attitude at the beginning. You need to take charge. View it as your meeting. Make it one that you will direct. This need not imply an aggressive stance. Just as a good chairperson may not speak first, loudest or longest, you can be in charge without making the other person feel overpowered. So, take the initiative and aim to *run* the kind of meeting *you* want – yet that the other person will find appropriate or like.

The first task is to get their attention, to make them concentrate on the issue at hand. You will never persuade anybody of anything if they are not concentrating on, and thus appreciating, what is involved. Imagine what they are thinking – 'Is this going to be interesting, useful or a waste of good time?' – and aim to make sure that their first reaction is as you would want it. Perhaps something like, 'This seems as if it will be useful. So far, so good. Let's see what they have to say.'

To create this impression it helps if you:

- appear well organized and prepared;
- suggest and agree an agenda that makes sense to you both;
- make clear how long the session will last;
- get down to business promptly.

Overall, if in the first moments you show interest in the other person and make it clear that they are important to the proceedings, this will certainly help. Even something as simple as a little flattery may help: 'Some of your good organization would help here, John, can you spare ten minutes to go through …?' Of course, not everyone is susceptible to this sort of thing – hold on: if you just said, 'That's right' to yourself, you have shown how useful this can be!

4.5.2   Finding out

With the meeting under way, the next stage is to find out something about the other person's perspective on the matter.

Such finding out is achieved by asking questions – and *listening* to the answers; here again you can usefully refer to earlier sections.

- **Questioning**: What to ask and how to put it may need some thought as you prepare. You need to phrase questions clearly and it is useful to use three levels of questioning (that is *closed* questions, *open* questions and *probing* questions). To extend the example: ask your colleague if there should be a presentation rehearsal and the yes-or-no answer tells you little. Follow up a yes answer by asking why they think it is necessary (an open question) and you will learn more – 'I'm really a bit nervous about the whole thing' – and more questions can then fill in the detail.

Panel 4.2 Example (continued)

Returning to the example of a presentation rehearsal, it may be useful to know whether your colleague:

- wants to rehearse;
- needs to do so;
- sees it as being done at any particular moment;
- envisages it as taking a particular amount of time ...

...and what they believe the presentation should achieve and how it might be done, and so on. Having some knowledge of this kind of thinking, and perhaps of their presentational abilities, shows you something about the job of persuasion to be done. This may range from a major battle (they do not want to do it at all), to a near-meeting of minds (you both see the need, but you are going to have to persuade them to give up longer for it than they envisage).

It is important here that people appreciate what is happening. Clever questioning may provide you with a useful picture, but this needs to be seen to be the case. You will persuade more certainly if the other person *knows* that you understand their position.

- **Listening**: Sounds obvious (but the details of making it work are set out in Chapter 3). Certainly, you should make no mistake: finding out can give you information that becomes the basis of successful persuasion. So, if the other party lets slip that their boss has said this had 'better go well', then later you might use that as part of your argument: 'Given what your boss said about it, perhaps the time we spend beforehand could be a little longer.'

Next, with some information to hand you can begin to put over your case.

4.5.3 The power of persuasion

The dictionary says of the word persuasion: to cause (a person) to believe or do something by reasoning with them. Fine, but the question is *how* to do this. To be persuasive, a case must be *understandable*, *attractive* and *credible*. Consider these in turn.

4.5.3.1 Creating understanding

A good deal has been said in other chapters about the need for clear communication. The point here is more than simply avoiding misunderstandings. People like clarity of explanation and ease of understanding. Spending five minutes going round the houses about something, only to have light dawn at the last moment in a way that gets the person thinking, 'Why ever didn't you say that to begin with?', hardly builds your credibility.

When people find something they expect to be difficult to understand easy, they like it. A powerful description, especially one that puts things in terms the other person can identify with, can strengthen a case disproportionately. Care is sensible here. Avoid inappropriate use of jargon: it is useful shorthand only when both parties have the same level of understanding of the terminology involved. You only have to think about computers to observe the problem. So, always:

- *think* about explanations and descriptions, try them out and be sure they work;
- aim to make what you say *immediately and easily understood*;
- be *thorough and precise*, telling people enough detail to make the point and emphasizing the most relevant points;
- *match the level of technicality* you use with the other person (and avoid or explain jargon if it might confuse).

This is an area where you can score some points. Think about the structure and sequence of what you say and how it breaks down into subsections. Present a logical and organized case and signal what you aim to do in advance: 'It may be easiest if we go through this in stages. Let's consider the timing first, then the costs, and then how we need to organize implementation.' If such a start gets people nodding – 'That seems sensible' – then you will carry them with you to the next stage. Use as many layers of this as are necessary to keep things clear. For example, in the above example you might add, 'Timing implies when we will do things and how long it will take. Let's discuss duration first, and then it should be easier to see when things can be fitted in.'

Already what you achieve in this respect can begin to put some convincing pluses on the positive side of your balance.

4.5.3.2   Making the case attractive

This part of the argument has to set out the core pluses of the case, painting a picture of why agreement should follow. You get your own way when people see what something does for, or means to, them. How this is done is largely a question of giving the argument a focus of what, in sales jargon, are called *benefits*, rather than *features*.

Benefits are what something does for or means to someone.
Features are simply factual points about it.

The spellchecker on a computer is a feature. Being able to produce an accurate manuscript quickly and easily, the time and effort saved and the avoidance of material being returned for correction (by a boss, say, or customer) are all benefits. They are things the feature – the spellchecker – allows to happen for me. Features act to produce benefits.

The sequence here is important. Just tell people everything about a suggestion in terms of its features and their response may well be to say (or think), 'So what?' Start by discovering what they want, then show them that what you are suggesting provides that and then the feature may reinforce the argument.

Panel 4.3   Example (continued)

Let's return again to the example. Say, 'A rehearsal will take only an hour' (the duration is a feature) and it may leave someone cold or get them saying, '*How long?*' in horror at what they see as a long time. Get them agreeing (a) that the presentation must go well ('Yes, it must'), (b) that there is a great deal to gain from it ('Right!'), and (c) that there is a possibility of two presenters falling over each other's feet unless there is a rehearsal ('Could be'), then talking of the 'ability of the rehearsal to increase the chances of success' (which is what it will do and is therefore a *benefit*) makes much better sense.

The principle described here is important. By thinking about the elements of the case in this way and, as you do so, adding to each point the thought, '…which means that …', you can tease out the most powerful description. This is a short chapter (its being short is a feature), *which means that* it does not contain many words; *which means that* it does not take too long to read; *which means that* you can apply any lessons you learn from it fast; *which means that* you may be able to get your way about something you have to discuss tomorrow. All this analysis moves the case more and more towards *something it will do for you* (a benefit). If you were addressing an individual and knew they wanted to raise the matter of their salary being increased tomorrow, then the case can be personalized and the benefit described made specific.

The task is therefore to make a clear case, to emphasize aspects of the case that have a positive effect on the other person and to make sure there are sufficient, and sufficiently powerful, pluses to add up to an agreeable proposition.

But there is a further element to making a persuasive case. It needs to be *credible*.

4.5.3.3   Adding credibility

Because of the inherent suspicion that tends to exist when selling or persuasion is in evidence, people's reaction to your saying that something is a good course of action to adopt may simply be to say, 'You *would* say that, wouldn't you?' Your say-so is not enough. They want more. Credibility is added to your case by your offering evidence, other than your opinion, that the case really is sound. The salesperson selling a car who says, 'The Automobile Association test results show it does 45 miles per gallon' is putting a reliable source ahead of the figure they want to quote, and boosting the weight it adds to the argument.

Such credibility can be added in many ways, for example by:

- quoting *past experience*: 'The project approach is very like ___, and that worked well';
- involving the *support of others* (a person or organization): 'The training manager says a rehearsal would be useful' (when the other party respects the person referred to);
- quoting *measurement of results*: '50 per cent of these kinds of presentation end without securing agreement; let's make this one of the successful ones';

- mentioning any *guarantees, tests or standards* that are met;
- invoking *quantity that reinforces the case*: 'Several departments work this way already'; 'Hundreds of people use it.'

It is worth thinking both about the need for proof and how strong that need may be, and thus what evidence can be used in support of your argument before exposing any case to others.

A final point here: remember that a person's perspective on something may not be solely their own. Someone may react with their employer or department, their boss, their family, or their staff in mind. Equally, they may react positively for reasons of common good, because their helping you will help make the department you both work for more efficient perhaps; or, very personally, they want to be seen to be involved in something, or you promise them a drink in the pub.

4.6   Feedback

At this point – arranging your message to ensure it combines being understandable, attractive and convincing – you may feel you are making a powerful case, but it is dangerous to assume so. You need some feedback, and thus it is important to include obtaining it in your approach. It is very easy to find that your confidence in a well-planned argument makes you forge ahead without pause, only to find later that they were with you only up to the moment you said, 'Right, let's make a start!'

Feedback can be obtained in two main ways.

(i) **Observation**: Look and listen. Do they look interested or are they tapping their fingers impatiently, eager to butt in, or gazing out of the window in boredom? Do they sound interested? Watch for remarks such as, 'That's interesting,' 'I see' and 'Fine' and for phrases that imply agreement, e.g. 'OK,' 'Should work' and 'Why not?'

(ii) **Ask questions**: To check understanding, e.g. 'Is that clear?', 'So far, so good?' To check their appreciation of benefits: 'Do you agree that would simplify things?' To check their reaction to features: 'How does an hour's meeting sound?' Or to check their perspective: 'You did say it would be best for you before the 10th?'

Responses of all sorts will clarify the picture as you go, helping you adjust your approach if necessary, and allowing you to focus on those areas that appear most readily to act as foundations of agreement.

So far, so good. If you make a powerful case you might move straight to agreement. Might. More often people have, in part at least, negative responses – they see a balance with some minuses as well as pluses. And maybe they see too many minuses to allow agreement. They object, in principle or to particulars; they confirm the old rule that there is always a but.

4.7   Securing agreement

The better you present your case – and the better it is directed towards the person in question – reflecting their situation, needs and views, the fewer *objections* you are likely to get, such as any reservations, either in someone's mind or voiced, that add weight to the negative side of the 'decision-making balance' and thus may lead to persuasion failing. So, although you are almost always going to get some, the first strategy is to reduce them by presenting a case that is 'on target' in this way.

4.8   Anticipating objections

The second point to bear in mind is that often it is not difficult to anticipate the nature of objections. If you know your fellow presenter in our example is busy, perhaps with a major project with pressing deadlines, then it should not be surprising if their first response is to find the time a rehearsal will take unacceptable. Anticipation does not mean that objections then become easy to deal with, but at least you have time to consider how best to handle them.

That said, sometimes the way in which routine objections come up surprises. Something you expect to be a major issue fails to materialize, something minor assumes giant proportions or something comes up late when you expect it early on; or vice versa.

Whatever objections may be about, they need handling. Ignored, unexplained or allowed to retain a major role in the balance, they can push the total balance into the negative – and the result is that you fail to get agreement.

4.9   Options for handling

The first thing about objections is to recognize that they are likely to occur and take a positive view of them. Think of them from the other person's point of view for a moment. They are trying to assess your proposition – weighing it up – and they think there are snags. They want you to take any point they raise seriously, not to reject it out of hand, which will seem unreasonable.

So consider some initial factors as rules.

- Regard objections as a sign of interest (after all, why would anyone bother to raise issues about something they had already decided to reject?).
- Anticipate and, perhaps, pre-empt them (especially regularly raised issues).
- Never allow arguments to develop (especially not of the 'yes, it is; no, it's not' variety).
- Remember that a well-handled objection may strengthen your case.

Thus the first response to an objection being voiced should be not a violent denial, but an acknowledgement. This may be only a few words – 'That's certainly something we need to consider'; 'Fair point, let me show you how we get over that' – but it is an important preliminary. It acts to:

- indicate that you believe there is a point to be answered;
- show you are not going to argue unconstructively;
- make it clear that your response is likely to be considered and serious;

- give you a moment to think (which you may need!);
- clarify what is really meant (if it is not clear what is being said, or why, a question may be a valuable preliminary to answering).

A well-handled acknowledgement sets up the situation, allowing you to proceed with the other person paying attention and prepared to listen. But you cannot leave things hanging long: you need to move on to an answer. The ways in which negative factors can be handled are mechanistically straightforward. There are only four different options, though all of them may need to be used in concert with stressing, or stressing again, things on the plus side of the balance. The four options are these.

1. **Remove them**: The first option is to remove the objection, to persuade the person that it is not actually a negative factor. Often objections arise out of sheer confusion. For example, if it is said, 'I don't have time for a full rehearsal!' this may be based on an overestimate of how long it will take. Tell them that what you have in mind is an hour or so, and not the whole morning they envisaged, and the objection evaporates.
2. **Reduce them**: You can act to show that, although there is a negative element to the case, it is a minor matter: 'Getting this presentation right is so important. It'll take a moment, certainly, but surely an hour or so is worthwhile.'
3. **Turn them into a plus**: Here you take what seems like a negative factor and show that it is, in fact, the opposite: 'Rehearsal seems elaborate and it'll take an hour or so, but we both have to do some individual preparation. Rehearsal will halve that time and ensure the presentation goes well.'
4. **Agree**: The last option, and one that the facts sometimes make necessary, is to agree that an objection raised is a snag: 'You're right: it *is* time-consuming, but this presentation has to go well and there's no other option.'

Because the options for dealing with the matter are only four, the process is manageable and it should not be difficult to keep in mind during a conversation and decide, as something is raised, how to proceed.

In every case keep the vision of the whole balance in mind. The job is *not* to remove every single minus from the negative side (there may well *be* some snags and this is simply not possible): it is to preserve the overall configuration of the balance you have created in the other person's mind.

4.10   Excuses

Let us be honest, sometimes people disguise their reasons for not acting as we wish. They say, 'It'll take too long' or 'The cost's too great' or whatever when they are simply being stubborn. In this case you need to try to recognize what is an excuse and what is not. A long justification of time or cost will achieve nothing if that is only a disguise for the real reason. For example, say your fellow presenter said they did not like the thought of presenting with you. Maybe what they are saying is that they are not very confident of their presentational skills and do not want you to witness them.

Suspect something like this is going on and the only way forward is to ask questions, and perhaps to drive things out in the open: 'Be honest, that's not really an issue. Why do you *really* object?'

4.11   Reaching a conclusion

Once your case is explained and all objections have been raised and dealt with, what next? It is easy to leave things without getting a decision. 'Has that given you all the information you need to make a decision?' is polite, but may just prompt someone to agree that it has, and try to drop the conversation for the moment. You have to *close*, i.e. actively *ask* for someone's agreement in order to obtain a commitment or prompt an action and lead towards a conclusion of the conversation. If you have made a good case, this is not so much part of the persuasion. Rather, it is only aiming to convert interest and agreement into action. You simply need to ask, and there are a variety of ways of doing this.

- Just ask ('Shall we put a time in our diary?').
- Tell them. You may not have the authority to instruct them, but make it *sound* like an order ('Put something in your diary').
- Suggest why it is a good idea to commit now rather than later ('Let's set a date now, while we can find a mutually convenient time that doesn't disrupt anything else too much').
- Suggest why it is a bad idea to leave it ('Unless we set a date now, we'll never find a convenient time').
- Suggest alternatives, positive alternatives where agreement to either one gives you your own way ('So, shall we clear an hour for this or make it two?'). And repeat as necessary ('So, an hour it is, then. This week or next?').
- Assume agreement and phrase the request accordingly ('Fine, we seem to be agreed. Let's get our diaries out and schedule a time').

Sometimes, after what has been a long, complex discussion, it may be useful to summarize as you conclude, touching once more on the main advantages of the action to which you seek agreement.

4.12   What next? – follow-up action

At this point you may get agreement. Or not. We all need to be realistic about this. No one *always* gets their own way. What you want is a good strike rate as it were, and going about it the right way helps achieve this. But there are other considerations. You need to do the following.

- **Deal with indecision**: If someone says they want to think about it, always agree: 'OK, it's an important decision, I can see you want to be sure.' But then try to find out why this is necessary: 'But are there any particular things you're not sure about?' Often something will be identified, maybe several areas (try to get a list), in which case assume the conversation will continue, revisit these areas and then move back to asking for a commitment as if there had been no hitch. Often this will then get a decision.

- **Follow up persistently**: If things have to conclude on a 'leave it with me' basis, agree the timing, keep the initiative: 'Right, I'll give you a call on Friday and we'll try to finalize things then.' Always take the follow-up action as planned – and go on doing it as long as necessary. Take all delays at face value. If a secretary says someone is in a meeting, assume they are, and move on: 'When would be a good time to contact them tomorrow?' Giving up can simply see agreement going by default for no good reason other than your lack of persistence.

4.13   Summary

Those most likely to get their own way do not charge at the process like bulls at a gate. They treat people with respect, trying to understand their point of view and using that in the argument. They may be assertive, persistent and thorough, but they are not aggressive (that may work once or twice, but is likely to cause resentment and make things more difficult in the long term).

They go about things in a systematic way, after spending some time preparing, and are patient yet insistent. People may agree to things for all sorts of reasons. But what persuades most readily is a reasoned case, something designed to make the particular individual (or individuals) respond positively because it is designed to persuade *them*. To make it do so you will:

- be prepared;
- act with consciousness of the psychology involved;
- get off to a good start;
- ask questions to establish others' needs or interest in the issues;
- structure your approach around the *other* person;
- take one point at a time;
- proceed in a logical order;
- talk benefits: tell them what your proposition will do for, or mean to, *them*;
- always be clear and descriptive;
- provide proof to back up your argument if necessary;
- check progress by obtaining feedback and keeping the conversation two-way;
- ensure that you appear (better still are!) confident about the whole exchange.

Practice makes perfect, though persuasion is as much an art as a science. But, given that, the techniques certainly help and understanding them is the first step towards drawing on them and deploying them appropriately to help you get your way more often and more certainly.

Chapter 5
Meetings: making them work

It was J.K. Galbraith who said, 'Meetings are indispensable when you don't want to do anything.' Then it is said that the ideal meeting involves two people – with one absent.

If there is one thing in business life that is a mixed blessing, then it is surely meetings; and technical people are likely to have little patience with them. Yet so much time is spent in them (and that means they cost), and they *are* an important part of organizational communications, consultation, debate and decision-making. We all need them. Or certainly we need some of them. But we must get the most from them, and we do not need too many, or those that are longer than necessary or, above all, those that are unconstructive. What is more, good, effective meetings do not just happen. If it is assumed that some deep law of meetings means you must put up with the bad ones in order to get an occasional good one thrown in, then nothing will be done to create a culture of effective meetings. Everyone in an organization needs to work actively at it. Everybody's role is important, whether they are running a meeting or attending one.

5.1   They have their uses

Whatever the meeting, large or small, formal or informal, long or short, if it is planned, considered and conducted with an eye on how it can be made to go well, it can be made to work.

As has been said, we all need some meetings, and their role and importance can vary. Meetings are simply a form of communication; they can:

- inform;
- analyse and solve problems;
- discuss and exchange views;
- inspire and motivate;
- counsel and reconcile conflict;
- obtain opinion and feedback;
- persuade;
- train and develop;
- reinforce the status quo;
- instigate change in knowledge, skills or attitudes.

You can no doubt add to the list, and will recognize meetings that aim to do several of these things. The key role is surely most often to prompt change (there is no point in having a meeting if everything is going to remain the same), and for that to happen decisions must be made. And to do that any meeting has to be construcctive and put people in a position where good decisions can prompt appropriate action.

It is also worth noting that good meetings are not just useful: most people *want* meetings. Having too few can be as big a mistake as having too many. Why do people want them? For various reasons, they believe meetings:

- keep people informed and up to date;
- provide a chance to be heard;
- create involvement with others;
- are useful social gatherings;
- allow cross-functional contact;
- provide personal visibility and public-relations opportunities for the individual;
- can broaden experience and prompt learning.

And more. And they are right. Meetings are potentially useful. Indeed, the progress of an organization can, in a sense, be certain only if meetings are held and if those meetings do go well.

5.2   Unnecessary costs

On the other hand, poor meetings can be costly; and why is it that in some organizations such meetings are just put up with, with people often exiting them muttering 'What a waste of time!' yet nothing is being done about it? The dangers here include meetings that:

- waste time;
- waste money;
- divert attention so that more important tasks suffer;
- slow progress and delay action;
- are divisive;
- lower morale;
- are a platform for the talkative and disruptive;
- breed office politics;
- create muddle, at worst chaos.

So, what people could be doing if they were not in a meeting must always be considered and no one in an organization should take meetings lightly – 'It's just a meeting.' They need to be productive and useful.

5.3   Before meetings take place

If a meeting is to be truly successful, then ensuring such success cannot start as the meeting starts – the 'I think we're all here, what shall we deal with first?' school of meeting organization. Making it work starts *before* the meeting – sometimes some time before.

First, ask some basic questions. For example consider the following.

- Is a meeting really necessary?
- Should it be a regular meeting? (Think very carefully about this one, for once a meeting is designated as the weekly (monthly or whatever) such-and-such meeting it can become a routine that is difficult to break, and this is an especially easy way to waste time.)
- Who should attend? (And who should *not*?)

If you are clear in these respects, then you can proceed to those matters that must be considered ahead of the meeting. These include the following.

- **Setting an agenda**: This is very important; no meeting will go as well if you simply make up the content as you get under way (notify the agenda in advance and give good notice of contributions required from others)
- **Timing**: Set a start time and a finishing time, then you can judge the way it is to be conducted alongside the duration and even put some rough timing to individual items to be dealt with. Respect the timing too: start on time and try to stick with the duration planned.
- **Objectives**: Always set a clear objective so that you can answer the question *why* a meeting is being held (and the answer should never be – *because it is a month since the last one!*).
- **Preparing yourself**: Read all necessary papers, check all necessary details and think about how you will handle both your own contribution and those of other people.
- **Insisting others prepare also**: This may mean instilling habits (if you pause to go through something that should have been studied before the meeting, then you show that reading beforehand is not really necessary).
- **People**: Who should be there (or not) and what roles individuals should have.
- **Environment**: A meeting will go much more smoothly if people attending are comfortable and if there are no interruptions (so organize switching the coffee pot on and the phones off before you start).

5.3.1   The agenda

Most of the points above are common-sense ones. It is, however, worth expanding on the key issue of the agenda – that is the document setting out topics for discussion at a meeting, the sequence in which they will be dealt with and administrative information such as timing and location.

Every meeting needs an agenda. In most cases this needs to be in writing and is best circulated in good time ahead of the meeting; it is intended to help shape and control the meeting that follows. It should achieve the following.

- Specify any formalities (must such things as apologies for absence be noted, for instance?).

- Pick up and link points from any previous meeting(s) to ensure continuity.
- Give people the opportunity to make agenda suggestions.
- Specify who will lead or contribute on each item.
- Help individuals prepare for their participation.
- List and order the items for discussion or review. This is something that may need to represent the logical order of the topics, the difficulty they pose, the time they will likely occupy, and participants' views. Some compromise is normally necessary here.
- Deal with administration: when and where the meeting will be held and any other arrangements that need to be noted in advance.

Once a draft agenda is prepared, it should be checked against any practical issues before being circulated. For instance, will the meeting it suggests fit in the time available, is there time for any necessary preparation, will one item overpower others (necessitating two meetings perhaps)? And so on.

Then, at the appointed hour, someone must take charge and make the meeting go well.

5.4   The role of leading a meeting

Even a simple meeting needs someone in the chair. That does *not* imply that whoever is 'in the chair' should be most senior, should do most of the talking or even lead the talking, or that they need to be formally called Chairperson – but they should be responsible for *directing* the meeting.

An effectively conducted chairing role can ensure a well-directed meeting, which, in turn, can mean:

- the meeting will better *focus on its objectives*;
- discussion can be kept *more constructive*;
- a *thorough review* can be assured before what may otherwise be ad hoc decisions are taken;
- all sides of the argument or case can be *reflected and balanced*;
- proceedings can be kept *businesslike and less argumentative* (even when dealing with contentious issues).

As we see, all the results of effective chairing are positive and likely to help make for an effective meeting. Put succinctly, a good chairperson will lead the meeting, handle the discussion and act to see that objectives are met, promptly, efficiently and effectively and without wasting time.

Some of what must be done is simple and much is common sense; the *whole* of the role is important. We will start by referring to two key rules that any chairperson must stick to (and that any group of people meeting should respect). They are, very simply:

1. only one person may talk at a time;
2. the chairperson decides who (should this be necessary).

Already this should begin to make you think about the qualities of the person who will make a good chairperson. The right choice must be made: the person in the chair must be able to execute their task effectively, must be respected by the participants (who must accept first that a chairperson is necessary). The next few points set out what the chair must do.

5.4.1   The meeting leader's responsibilities

The list that follows illustrates the range and nature of the tasks involved. It also shows clearly that there are skills involved, perhaps skills that must be studied, learned and practised. Whoever is leading the meeting must be able to achieve the following.

- *Command the respect* of those attending (and, if they do not know them, then such respect must be won rapidly by the way they are seen to operate).
- *Do their homework* and come prepared, i.e. having read any relevant documents and taken any other action necessary to help them take charge. They should also encourage others to do the same, as good preparation makes for more considered and succinct contributions to the meeting.
- *Be on time.*
- *Start on time.*
- Ensure any *administrative matters are organized* and will be taken care of appropriately (e.g. refreshments, taking minutes).
- Start on the *right note* and lead into the agenda.
- *Introduce people* if necessary (and certainly know who's who – name cards can help everyone at some meetings).
- Set, and keep, the *rules.*
- *Control the discussion*, and do so in light of the different kinds of people who may be present: the talkative, the strident etc.
- *Encourage contributions* where appropriate or necessary.
- *Ask questions* to clarify where necessary. It is important to query anything unclear and do so at once. This can save time and argument, whereas, if the meeting runs on with something being misinterpreted, it will become a muddle and take longer to reach any conclusion.
- Ensure everyone *has their say.*
- Act to keep the discussion *to the point.*
- *Listen* (as in LISTEN): if the chair has missed things, then the chances of the meeting proceeding smoothly are low and it may deteriorate into 'But you said ...' arguments).
- *Watch the clock.* Remind others to do the same and manage the timing and time pressure.
- *Summarize*, clearly and succinctly; something that must usually be done regularly.
- Cope with upsets, outbursts and emotion.
- Provide the *final word* – summarizing and bringing matters to a conclusion. Similarly, link to any final administrative detail, such as setting the date for the next action or further meeting.

- See (afterwards) to any *follow-up action*. This may be especially important when there is a series of meetings, because people may promise something at one and turn up at the next having done little or nothing.

And all this must be done with patience, goodwill, good humour and respect for both all those present (and maybe others) and for the objectives of the meeting.

5.4.2   *The conduct of the meeting*

Now let us turn to a number of points worth investigating in more detail.

5.4.2.1   Get off to a good start

The best meetings start well, continue well and end well. A good start helps set the scene, and this too is the responsibility of whoever is in the chair. It works best to start the meeting in a way that:

- is positive;
- makes its purpose (and procedure) clear;
- establishes the chair's authority and right to be in charge;
- creates the right atmosphere (which may differ if it is to prompt creative thinking or, say, detailed analysis of figures);
- generates interest and enthusiasm for what is to come (yes, even if it is seen as a tedious regular review);
- is immediately perceived as businesslike.

It may also help if the chair involves others early on, rather than beginning with a lengthy monologue – which takes us to the next point.

5.4.2.2   Prompt discussion

Of course, there are meetings where *prompting* contributions is the least of the problems, but where contributions are wanted from *everyone* (or why are they there?), not just a few. So, to ensure you get adequate and representative discussion and that subsequent decisions are made on all the appropriate facts and information, you may need to prompt discussion.

For example, sometimes there are specific reasons why meeting participants hold back, for example:

- fear of rejection;
- pressure of other, more senior or more powerful, people;
- lack of preparation;
- incomplete understanding of what has gone before.

Or, indeed, it may simply be a lack of encouragement to make contributions. A good chairperson will ask for views and do so in a way that prompts open, considered comments.

But note that it is sometimes easy to skew comments (wittingly or not) by the tone or manner with which comments are called for. For instance, a senior manager is unlikely to encourage creative suggestions if they field their own thought first: 'I'm sure this is an excellent idea – what do you think?' So, do not lead.

The other trick is to ensure you have the measure of different individuals, drawing in, say, the more reticent and acting to keep the overbearing in check. And remember that people may have a variety of motives for the line they take at meetings, so some will be more businesslike than others.

Much comment-prompting will come through questions and the way this is done is important.

Questions must of course be clear. Remember also that there are two main kinds of question, open and closed, with open questions more likely to prompt discussion (see 4.5.2 in Chapter 4). The circumstances will affect how questions are best asked. Discussion can be prompted around the meeting using more complex means, primarily in the following six ways.

1. **Overhead questions**: These are put to the meeting as a whole, left for whoever picks them up and are useful for starting discussion.
2. **Overhead/directed**: These are put to the whole meeting (as in 1 above) and either followed immediately by the same question to an individual, or after a pause as a way of overcoming lack of response: 'Now, what do we all think about this? [Pause] David?'
3. **Direct to an individual**: Direct to an individual without preliminaries; useful to get an individual reaction or check understanding.
4. **Rhetorical**: A question demanding no answer can still be a good way to make a point or prompt thinking and the chairperson can provide a response if they wish. 'Useful?'
5. **Redirected**: This presents a question asked of the chair straight back to the meeting either as an overhead or direct question: 'Good question. What do we all think? David?'
6. **Development question**: This really gets discussion going. It builds on the answer to an earlier question and moves it round the meeting: 'So, Mary thinks it'll take too long. Are there any other problems?'

Prompting discussion is as important as control. It is the only way of making sure the meeting is well balanced and takes in all required points of view. If decisions are made in the absence of this, someone may be back to you later saying something like, 'This is not really acceptable. My department never really got a chance to make their case.'

Because of this it may sometimes be necessary to persevere in order to get all the desired comment the meeting needs. Ways have to be found to achieve this; here are two examples.

• **Ask again**: As simple as that. Rephrase the question (perhaps it was not understood originally) and ensure the point is clear and that people know a comment *is* required.
• **Use silence**: The trouble is that silence can be embarrassing. But even a short silence to make it clear you will wait for an answer may be sufficient to get

someone speaking. So do not rush on; after all, maybe the point deserves a moment's thought.

5.4.2.3   Concentrate

A good serious meeting demands concentration. It is the job of whoever is in the chair to assist achieving this in them and in others. Beware interruptions! Organize to deal with messages, mobile telephones or simply the refreshments arriving mid-meeting, as all will delay proceedings and ensure concentration is lost.

It helps, therefore, if:

- rules are laid down about messages;
- breaks are organized (for longer sessions) so that people know how they can deal with messages etc., and are sufficient to maintain concentration;
- refreshments are organized in advance (or after the meeting);
- others outside the meeting (including switchboard operators and secretaries) are briefed as to how matters should be handled (it is as bad for a key customer, say, to be told, 'Sorry, they're in a meeting' as it is for a meeting to be interrupted, perhaps worse, so deciding the priorities is important);
- in the case of unforeseen interruptions, you do not compete with them while people's attention is elsewhere: wait, deal with them, and then continue, recapping if necessary.

Concentration is vital, and, of course needs to be focused on the right things within the meeting. Do not be sidetracked; beware of digressions; beware of running a meeting within a meeting – true, sometimes you *will* unearth separate issues that are worth noting to pursue or investigate on some other occasion, so note them but do not let yourself become distracted by them.

5.4.2.4   Keep order

Sometimes even the best-planned and best-organized meeting gets out of hand. So here it is worth noting some further key rules for the chair.

- Never get upset or emotional yourself.
- Pick on one element of what is being expressed and try to isolate and deal with that without heat and to reduce the overall temperature.
- Agree (at least with the sentiments) before regrouping: 'You're right, this *is* a damned difficult issue, and emotions are bound to run high. Now let's take one thing at a time.'

If these approaches do not work you may have to take more drastic action, for example:

- call for a few minutes' complete silence before attempting to move on;
- call a short break, and insist it be taken without discussion continuing;
- put the problem item on one side until later (though be sure to specify how and when it will be dealt with and then make sure you do what was agreed);
- abandon the meeting until another time.

This last option is clearly a last resort, but ultimately may be better than allowing disorder to continue. Usually a firm stand made as soon as any sort of unrest occurs will meet the problem head on and deal with it. Whatever happens, as it says in *The Hitchhikers' Guide to the Galaxy*, 'Don't panic!'

It would be wrong to give the impression that chairing meetings is all drama – although, if discussion never got heated, what would that signify? So let us end this section with something more constructive.

5.4.2.5   Sparking creativity

It is said that managers are not paid to have all the good ideas that are needed to keep their department or whatever running effectively, but they *are* paid to *make sure there are sufficient ideas to keep it ahead*. As a result, many meetings need to be creative. Two heads really can be better than one; yet also new ideas can prompt a negative cycle all too easily. Such discussions can spiral into a tit-for-tat of 'Your idea's no good' or 'My idea's better' – scoring points taking precedence over giving the new thinking a chance.

This too is something the chair has to deal with, fostering creative thinking and open-mindedness and ensuring that instant negative reactions are not the order of the day. The chairperson must, therefore:

- actively stimulate *creative thinking* (saying this is part of the meeting and ruling against instant rejection of ideas without consideration);
- contribute *new ideas* themselves or steer the discussion in new or unusual directions;
- find *new ways* of looking at things;
- consider *novel approaches* and give them a chance;
- aim to *solve problems*, not tread familiar pathways.

Some groups who meet regularly get better and better at this, but this does not usually happen spontaneously. More often it is the result of someone putting together the right team and prompting them to think along certain lines and, above all, remain always open minded.

5.5   The individual attendee's responsibility

It is certainly vital for a meeting to be well led, but everyone attending must play their part. Anyone, ahead of attending any meeting, should ask themselves:

1. What can I contribute and how?
2. What can I get from the meeting?

This may lead into some preparatory action to get up to speed on the issues or whatever. Never just walk into a meeting resolving only to *see what happens*.

It is said that you get out of things what you put in, and this is certainly true of meetings. The key things that make for effective participation in meetings are the following.

- **Sound preparation**: You have to know what is to be discussed, be ready for it and have done your homework.
- **Effective communication**: You have to judge well what you want to say and how you want to say it.
- **Well-handled discussion**: You have not only to be ready to make planned contributions, but also to think quickly in order to respond to the hurly-burly of debate (or argument!). Both may be necessary to make your case.

One thing that is worth bearing in mind is that a meeting is a public forum; you are on show. There may be people unknown to you, senior people or important people, and appearance is a factor. I would not presume to tell you how to dress, but you do need to appear organized and as if you have some clout. If you rush in late, unkempt, and clutching a collapsing pile of papers you will certainly not give the right impression. Enough said. In some forums, seating may be important, too, and you need to avoid being squeezed into some corner barely visible to the chair.

5.5.1   *Maximizing the effectiveness of individual participation*

Now let us take the three key issues in turn.

5.5.1.1   Sound preparation

Preparation has already been mentioned as a prerequisite to success; meetings are no exception. Specifically here, you need to do the following.

- **Read everything necessary in advance**: This may include past minutes, agendas and documents of all sorts, some circulated in advance, others simply good background. Sometimes other research may be necessary.
- **Annotate any relevant documents and make your own notes on them**: It is essential to be able to check facts fast during a meeting, rather than fumbling and beginning lamely, 'I know it's here somewhere.'
- **Note who else will attend**: Are they allies or not, and is any liaison useful before the meeting starts?
- **Plan questions**: Consider what you will ask and of whom, and when and how you will put it.
- **Prepare any formal contribution**: Think through things much as you would plan a presentation; certainly, do not assume an informal meeting makes this unnecessary.

Bring notes that any of the above creates to the meeting and keep them accessible. Remember that meetings are mixed in every sense: people, ideas, motivations – even the most sensible idea may need selling and, as meetings can sometimes be daunting occasions, preparation helps reduce any nerves you may feel and prevent 'stage fright'.

5.5.1.2   Effective communication

All the basic rules apply. Here we review some factors specific to meetings. So …

- **Stick to the rules**: Do not monopolize the conversation, constantly interrupt, become emotional or argumentative, make it difficult to keep to time, appear unprepared, undisciplined or a troublemaker – and stick to the point.
- **Break the rules**: On the other hand, there is room for more dramatic impact (though as an exception), so you might want – *very carefully* – to stage: a dramatic outburst (a fist banged on the table, for instance), a display of emotion, some humour or apparent temper – even an organized interruption has been known to create advantage
- **Stick to the structure**: Communicate in an organized fashion: introduce what you want to say; set out the issues (maybe costs, timing, staffing, whatever), putting points for and against and dealing with both main issues and other implications; be prepared to debate points and answer questions; and summarize and pull together your argument.
- **Get your facts right**: Much time is wasted in meetings because of imprecise information (back to preparation and checking). Worse, one incorrect fact can cast doubt on everything you say. So, ensure you are: explicit (stating things plainly and not obscuring them with irrelevances), accurate and precise (picking just the right information to make your point).
- **Always observe others**: Watch reactions, listen to everything and try to put yourself in others' shoes. For instance, if someone seems negative ask yourself why that might be.

In addition (and see Chapter 4 for this), remember that making any case successfully demands that you make it understandable, attractive and credible; it is often necessary at meetings not just to tell people something, but to sell it to them.

5.5.2 Well-handled discussion

Much here is dependent on the chair, but an individual can influence things too. You will find you are able to perform best if you:

- *remain alert* and concentrate on everything that goes on (not appearing to be uninterested in parts of the meeting even if they are of less importance to you);
- *listen* – something now stressed more than once;
- *keep thinking* – you may have to fine-tune everything you planned in the light of events;
- remain *calm and collected*, whatever the provocation, and stick to your guns in a businesslike manner.

What about timing? You need to think about *when* exactly to speak: make a point too early, and it may be forgotten by the end of the meeting; leave it too late and the end of the meeting may come before you have a chance to have your say. So …

- Do not plan an input that will fit into only one spot. Remain flexible.
- Consider your comments, before and during a meeting, in context of what others are saying or might say.

- Play to your strengths: are you better at introducing something or adding weight to a case someone else initiates? Maybe you have to work in concert with others.
- Announce intentions: for example, saying, 'I have some thoughts on the costs of this; maybe I can reserve a word for when we discuss Item 3.'

Resolve never to leave a meeting saying to yourself. 'If only I'd said …' Start thinking ahead to your next one now.

5.6   After the meeting

The key thing after a meeting is that the actions decided upon should be implemented – as intended and on time; it is that which ultimately makes a meeting worthwhile. This may involve anything from a small piece of information being passed on, to the implementation of a major project. Whatever it is, however, it may well need prompting.

5.6.1   Are minutes necessary?

We do not want more paperwork in an organization than is necessary, so sometimes minutes are not necessary. But sometimes written follow-up is necessary, though this may range from a simple written note to full-blown minutes, so that there is an action reminder and a record. Three key reasons make minutes necessary. Minutes provide:

1. a *prompt to action*, reminding those who have taken on, or been given, tasks that they should do them, and do them on time – more so when several people must liaise to instigate action;
2. a *tangible link* to follow-up discussions or a further meeting; this can help ensure that points are reported or taken further – the classic 'Matters arising' item from minutes;
3. a *record* of what has occurred and particularly of what decisions have been made and what action decided upon; some records clearly need keeping longer than others, and this is dependent on the topic and import of what was the subject of the meeting.

So, minutes are to be prepared only when necessary, not as a routine, and when they are necessary they must be clear. So, minutes, or any kind of action note, must be kept practical and should be:

- **Accurate**: This may seem obvious, but care is necessary, as it is important if arguments are to be prevented in future.
- **Objective**: The job of the minute writer is to report what happened, not embellish it, especially not with their own view.
- **Succinct**: Unless they are manageable, minutes are likely to remain unread – brief, but incorporating all the essentials, is the rule.
- **Understandable**: Always important (as has been mentioned more than once in this book).
- **Businesslike**: Their key purpose is to spell out *what* action is expected, *by whom* and *when*. These key points should be **highlighted** in some way in the document.

So, minutes must: be attractive and easy to read; deal with formalities (apologies, minutes of the last meeting, matters arising, etc.); list the essence of the discussion and outcomes; be circulated promptly and regarded as important documents.

Meetings need not be a question of slavishly following rules. Whoever is in the chair can be as creative as they wish, finding ways to bring a meeting to life, while ensuring that it is effective and businesslike.

For example, some make a point of switching from the traditional format of *ending* with the ubiquitous 'Any other business' (AOB), which too often sees the meeting nosedive into a mess of bits and pieces, gripes and irrelevances, leaving people irritated when they should be saying, 'That was a useful session.' Better, perhaps, to *start* with AOB: a chairperson saying, 'Let's take ten minutes before we really get started to deal with any peripheral points and get those out of the way so that we can concentrate on the main business' is often appreciated – and creates the opportunity to finish without irritation.

5.7   Summary

It is vital that meetings be viewed seriously; it is never 'just a meeting'. Among the key issues are:

- scheduling only necessary meetings (and never letting them become an unthinking routine; especially regular meetings);
- never skimping preparation: meetings need a clear purpose, an organized agenda that respects the timings, and adequate time and instructions for everyone attending to get themselves ready for the event;
- effective control – a firm and thorough person in the chair and everyone behaving in an orderly way;
- participation – it is surely important that if everyone attending is present for good reason their view should be heard (part of the chairperson's task is to ensure this happens and to summarize and distil the contributions);
- that meetings lead to decision and action, so minutes and follow-up action must be punctiliously organized.

Overall, meetings must be constructive. If they are they will be stimulating (well, more stimulating than badly executed ones – no one pretends all meetings are fun). Good meetings certainly do not just happen, but they can be *made* to happen, and when they are the benefits can be considerable.

One last thing, something that can be usefully made into a rule: *always finish on a high note.*

Chapter 6
Presentations: speaking successfully 'on your feet'

How do you feel if you know you have to make a presentation? Confident? Apprehensive? Terrified? If it is apprehensive or even terrified, relax – you are normal. Most people share your feelings to some extent.

In fact, psychologists tell us that feelings about presenting or public speaking are almost wholly negative. What they call 'self-talk' prior to having to speak in public consists of a catalogue of intimations of disaster. We say to ourselves things such as, 'I can't do it'; 'I'm not ready'; 'The audience will hate it' (or turn against you); 'I'll dry up'; 'I won't have sufficient material'; 'I don't know how to put it'; 'I'll lose my place'; and, at worst, 'I'll die.'

What is more, if you stand up unprepared and ignorant of how to go about it, your presentation may well be a disaster. But it need not be. With some knowledge, some practice and having thought it through, there is no reason on earth why you should not make a good job of it. Here we review just how to go about it; but, before turning to that, first consider the nature of presentations.

6.1 The importance of presentations

Presentations are important.There can be a great deal hanging on them – a decision, an agreement, a sale – and they can affect financial results and reputations too.

All sorts of things can be involved. Presentations may be necessary:

- at an internal meeting;
- externally, to distributors or customers, perhaps;
- to a committee (or board);
- at a conference or business event;
- at a social event (anything from something in a business context, such as a retirement party, to a wedding).

And there are more, of course. You may have to speak to people you know, or those you do not know; to more senior people, difficult people or those who are younger or older than you. You may need to speak to ten, twenty or two hundred people; or more.

All events and all groups exhibit one similar characteristic: they judge you by how you present. An example makes this clear. Imagine that you have to announce some policy change. Let us assume it is something eminently sensible that should be accepted without trouble. But start, 'Now …um …what I want to say is …well, that is, basically …' and you may be well on the way to getting even the most innocuous message rejected.

It is a fact of life that something poorly presented can be poorly received, despite the sense of its content. People do not think, 'What an excellent idea! What a shame it wasn't put over better!' They think, 'What a rotten presentation! I bet the ideas were rotten, too.'

6.2   A significant opportunity

It is precisely because of this that presentations create a real opportunity. If you can do them well (and you can), you will positively differentiate yourself from others, who, from ignorance or lack of care, do an undistinguished or poor job.

It is worth quoting here a phrase used in a training film about making presentations, where a character describes presentations as being the 'business equivalent of an open goal'. Well put! This is not overstating the point and puts it in a memorable way (I quote from the excellent film *I Wasn't Prepared for That*, produced by Video Arts Limited).

So, motivation should not be in doubt here. There are few business skills more worth mastering than that of presentation. Without it, you not only *feel* exposed, you *are* exposed. The trouble is the ground does not mercifully open up and allow you to disappear along with your embarrassment. It is more likely that the result is much more real – no agreement, no commitment or the boss saying ominously, 'See me afterwards.'

There are good reasons for having fears; all can either be overcome or reduced to stop them overpowering your ability to work successfully. It may help to think of things as a balance. On one side there are things that can, unless dealt with, reduce your ability to make a good presentation. On the other, there are techniques that positively assist the process. The right attention to both sides improves your capability.

Much here is about the positive techniques. But let us start with a little more about possible difficulties, some of which are inherent to the process, and how to overcome them – if only to get the negative side out of the way first.

6.3   The hazards of being 'on your feet'

We communicate so much that we tend to take it for granted. Indeed, we regard much of it as easy, and people may well say of a presentation that they know they could so easily go through the content if sitting comfortably opposite just one other person.

The first rule here is easy. Ease of communication should never be taken for granted. As has been said, it needs thought, care and precision, and this is doubly so when you present formally. With most presentations you get only one crack at it, and often there is not the to-and-fro nature of conversation that establishes understanding.

This means every tiny detail matters. Presentations are inherently fragile. Small differences – an ill-chosen word or phrase, a hesitation, a misplaced emphasis – can all too easily act to dilute the impact sought.

At least communications problems constitute a tangible factor. If you resolve to take care, your communication will be better and understanding more certain. You can work at getting this right. Many of the elements reviewed as this chapter continues assist this process, but what about less tangible fears?

6.4 Presenters' nightmares

Whatever *you* fear will cause making a presentation to be more difficult, it is probable that *others* think the same. Asking groups on workshops I conduct on the subject about their worries usually produces a very similar list of factors.

The top ten, in no particular order, are as follows, listed here with some thoughts about overcoming them.

6.4.1 Butterflies in the stomach

If you are nervous, then you are likely to *feel* nervous. Without some apprehension, which can act to focus you on the job in hand, you would probably not do so well. Much of this feeling will fade as you get under way (and knowing this from practice helps), but you can help the process in a number of ways, for instance:

- taking some *deep breaths* before you start (nerves tend to make you breathe more shallowly and starve you of oxygen), and remember to breathe as you go along (running out of breath to the point of gasping is a surprisingly common fault);
- taking a *sip of water* just before you start;
- *not eating a heavy meal* before a presentation;
- *not eating nothing* (or else rumbles may join the butterflies);
- *not taking alcohol* (except possibly in extreme moderation) – it really does not help; at worst, it may persuade you that you can do something you cannot, and make matters worse as the truth dawns!

6.4.2 A dry mouth

This is easily cured. Take a sip of water. Never attempt to speak without a glass of water in front of you. Even if you do not touch it, knowing it is there is a comfort. And beware of fashionable fizzy water, which can have distracting side effects!

6.4.3 Not knowing what to do with your hands

The best solution is to give them something to do – hold the lectern or a pencil, make the occasional gesture – but then forget about them. *Thinking* about them as you proceed will make matters worse.

6.4.4   Not knowing how loud to speak

Just imagine you are speaking to the furthest person in the room (if they were the only one there you would have little problem judging it); better still test it beforehand.

6.4.5   A hostile reaction

The vast majority of groups want it to go well. They are disposed to be on your side. The only thing worse than knowing that you are not presenting well is being in the audience; think about it.

6.4.6   Not having sufficient material

This can be removed completely as a fear. If your presentation is well prepared (of which more anon) you will *know* there is the right amount.

6.4.7   Having too much material

As point 6.4.6. Enough said for the moment.

6.4.8   Losing your place

Also tied in with preparation (and something else we will review in detail), your notes should be organized specifically so that it is unlikely that you will lose your place (and so that you can find it easily should you do so).

6.4.9   Drying up

Why should this happen? Dry mouth? Take a sip of water. Lose your place? Organize it so that this does not happen. Or is it just nerves? Well, some of the factors already mentioned will help – so, too, will preparation. And, if it does happen, often it takes only a second to resume: 'There was another point here – ah, yes, the questions of …' The problem here can be psychological: it just *feels* as if you paused for ever.

6.4.10   Misjudging the timing

This is something else speakers' notes can help with specifically.

All that is necessary for so many such problems or thoughts is a practical response, something that acts certainly to remove or reduce the adverse effect. Thinking of it this way helps too. Try not to worry. No doom and gloom. It will be more likely to go well if you are sure it will – more so if you work at organizing so that every factor helps.

But few people can speak without thought. It was the author Mark Twain who said, 'It usually takes me three weeks to prepare a good impromptu speech.' Preparation is key to success and it is to that which we turn next.

6.5 Preparing to present

Imagine that you have a presentation to make. Maybe you have one to be done soon (if not, bear in mind that this is a task that most organizational jobs do not allow you to avoid). Few people will simply do nothing about it until the day and then get up and speak. So what do you do? Let us address some dangers first to lead into what is best practice here. What you might do is think of what you want to say first, then think of what follows – what you will say second, third and so on – and then write it down verbatim. Then, perhaps after some judicious amendment, you read it to the group you must address.

Wrong, wrong and wrong again.

This may sound logical, but contains the seeds of disaster. We will pick up some alternative approaches as we continue. As it is a straightforward factor to address, let us take the reading aspect first.

6.5.1 *Do not try to read verbatim*

Some people think, at least until they have more experience, that having every word down on paper and reading them out acts as a form of security blanket. After all, what can go wrong if you have everything, right down to the last comma, in black and white in front of you? Well, two things in particular.

First, you will find it is really very difficult to read anything smoothly, get all the emphasis exactly where it needs to be, and do so fluently and without stumbling. The actors who record novels, and other books, as audio works deserve their pay cheques: real skill is involved here.

Note: avoid at all costs the annoying mannerism favoured by many politicians of reading line by the line (especially from the teleprompter) and ignoring punctuation, so that all the pauses are at the line end: 'Good morning, ladies and ...gentlemen. I am here today to give you a ...clear insight into our policy on ...'

Most people speak very much better from notes that are an abbreviation of what they intend to say (we will return to just how much of an abbreviation and what you should have in front of you later). If you doubt this, just try it: read something out loud and see how it sounds; better still, record it and *hear* how it sounds.

In addition, certainly in a business context, you rarely need to be able to guarantee so exact a form of wording (there are exceptions, of course: a key definition or description may need to be word-perfect). It is usually more important to ensure that the emphasis, variety and pace are right – and that is what is so difficult to achieve when reading.

Second, preparation cannot be done in isolation. It links to two factors that are key in making an effective presentation. These are:

- purpose or objective – *why* exactly you are making the presentation;
- the view you take of your audience.

As many would say that the audience is the first key essential here, let us start with that.

6.5.2   Your audience

Everything is easier with a clear view of your audience. First, who are they? They may be people you know, men/women, expert or inexperienced about whatever topic you must address; there are many permutations here. Most important, however, are the *audience's expectations*: what do they want?

Put yourself in their place. Facing a presentation, what do you say to yourself? Most people anticipate its impact on them: will this be interesting, useful, long or short; what will this person be like; will I want to listen to them; how will what they have to say help me? Again, the permutations are many (though usually not too complicated to think through), but bearing audience viewpoint in mind is a major factor in ensuring a successful outcome.

Specifically, any audience *wants* you to:

- 'know your stuff';
- look the part;
- respect them, acknowledging their situation and their views;
- discover links between what you say and what they want from the talk;
- be given an adequate message, so that they understand and can weigh up whether they agree with what is said or not (this is especially important if you are going to suggest or demand action of them);
- make it 'right for them' (for example, in terms of level of technicality);
- hold their attention and interest throughout.

It is equally important to bear in mind what audiences *do not want,* which includes not being:

- confused;
- blinded with science, technicalities or jargon;
- lost in a convoluted structure (or because there is none);
- made to struggle to understand inappropriate language;
- made to stretch to relate what is said to their own circumstances;
- having to listen to someone who, by being ill prepared, shows no respect for the group.

A good presenter will always have empathy for the group they address; and it must be evident to them. Often this is something guided by prior knowledge. But it can, of course, vary, and you may well need to speak to groups you do not know well. Always find out what you can and make use of everything you do discover.

Some of what makes for the right approach here is an amalgam of the various techniques we will explore later. Some relate to immediate practical factors that every presenter does well to remember. For example, while I would not presume to tell you how to dress for a presentation, it bears thinking about. Professionalism is, at least in part, inferred from appearance. Personal organization too has a visual importance. You must not just be well organized, you must *look* well organized. Walking to the front, however confidently, is likely to be spoiled if you are clutching a bulging folder spilling papers in all directions, and start by saying, 'I'm sure I have the first slide here somewhere,' accompanied by fevered, fumbling attempts to find it.

6.5.3   Clear purpose

Rarely, if ever, are people asked just to 'talk about' something. The most crucial question any intending presenter can ask themselves is simply, 'Why is this presentation to be made?' If you can answer that clearly, it will be easier both to prepare and to present. Let us be clear here: *Objectives are not what you intend to* say: *they reflect what you intend to* achieve.

Apologies if this seems obvious, but I often observe presentations (often carefully prepared and brought to training workshops in the presenters' knowledge that they will be subject to critique), which are poor almost solely because they have no clear objectives. They rattle along reasonably well; but they do not *go* anywhere.

Objectives are therefore key – you can check back for details about setting them elsewhere.

6.5.4   How the group sees a presenter

Any business presenter must direct the group, must be in charge and must therefore look the part. There are some people who hold that the presenter should always wear a suit, or the equivalent in terms of formality for a woman. Certainly, appearance in this sense is important, though it should link to the culture and circumstances in which the presentation takes place. Similarly, you should normally stand up as opposed to sitting (there may be some sessions that can be run while sitting, but not many, and these are less our concern here). Not only does appearance differ, but also most people will actually perform in a different and more stimulating manner when standing – it somehow gets the adrenalin flowing. If standing is the chosen option, stand up straight, do not move about too much, and present an appearance of purposefulness.

The speaker is the expert and is, or should be, in charge, so appearance is a relevant factor.

6.5.5   How you see the audience

How you view the group is not, of course, simply a visual point: what is necessary is an *understanding* of the group, and the individuals in it, and an appreciation of their point of view and their way of seeing things. Presentations may well demand decisions of people: 'Do I agree?'; 'Can I see the relevance of this?'; 'Shall I agree with this point?' So it is necessary to understand the thinking process that takes place in the minds of those in the group in such circumstances. This is essentially the same as was set out in Chapter 4 in discussing persuasive communication. This will not be repeated here, but the essential approach stemming from it is this: that anyone making a presentation must not simply talk *at* their audience, but rather tailor their approach based on an *understanding of the audience's point of view*.

Now, as we remember all this, one of the dangers is at once apparent. This is that the other person's point of view can be neglected, or ignored, with the presenter focusing primarily, or only, on their own point of view. You should ensure that you do not become introspective, concerned with your own views or situation, but instead use and display enough empathy to come over as being constantly concerned about others' views. This sounds obvious, but it is all too easy to find your own perspective predominating, thus creating a dilution of effectiveness. Even the most important message has to earn a hearing, and this is achieved primarily through concentrating on what is important to the group. Nervousness of the actual process of presenting may compound this potential danger.

Next we turn to the structure of the presentation itself, and review how one goes through it.

Probably the most famous of all maxims about any kind of communication is the old saying, 'Tell 'em, tell 'em and tell 'em.' This can be stated more clearly as meaning that you should tell people what you are going to tell them, tell them, and then tell them what it was you told them. This sounds silly, perhaps, but compare it with something a little different: the way a good report is set out, for instance. There is an introduction, which says what it is that follows; there is the main body of the document, which goes progressively through the message; and the summary, which, well, summarizes or says what has been covered. The idea is straightforward, but, if it is ignored, messages may then go largely to waste.

So, practising to some degree what we preach, we will split the presentation into three sections, and look not only at how to make each effective, but also how to ensure that the three together make a satisfactory whole. First, a few more words about preparation.

6.5.6   *Before you speak*

Having seen there are three stages – which we review under the more businesslike headings of the beginning, the middle and the end – we start with another factor, which is either confusing or an example of an intriguing opening. In any case, it has been referred to before: preparation. It is that which creates your beginning, middle and end, and everything else along the way.

Here I wish unashamedly to emphasize the point. Preparation is important – remember Mark Twain. If he was half as good a speaker as he was a writer it makes a point. So, before we analyse a presentation, we need to think about how you put it together.

6.5.6.1   **Preparation: key tasks**

The key issues are as follows.

- Be able to answer the question, 'Why must this presentation be made?' And have a clear purpose in mind, one that reflects the audience and the effect you want to have on them.
- Decide what to say (and what not to say).
- Arrange things in a logical order.
- Think about how it will be put over (not just the pure content, but examples, anecdotes and any elements of humour).
- Prepare suitable notes as an aide-mémoire to have in front of you as you speak (but not, as has been said, to read verbatim).
- Anticipate reactions and questions and how you will deal with these.

All this must be done with a keen eye on how long there is for the presentation so that what you prepare fits (you may need to decide the time; or you may be told the duration or have to ask what is suitable).

6.5.6.2 A final check

A final look (perhaps after a break following preparation) is always valuable. This is also the time to consider rehearsal: talking it through to yourself, to a tape recorder or a friend or colleague, or going through a full-scale 'dress rehearsal'.

If you are speaking as part of a team, *always* make sure that speakers get together ahead of the event to rehearse, or at least discuss, both any possible overlaps and any necessary handover between speakers. You are seeking to create what appears to the audience to be a seamless transition between separate contributors.

Ask: is this the sort of event where rehearsal is necessary and, if so, how should it be done? Simple, unequivocal answer: yes, it is exactly where such is necessary and this is how should it be done: in a nutshell, thoroughly, sufficiently far ahead, and taking sufficient time for it.

Thereafter, depending on the nature of the presentation, it may be useful – or necessary – to spend more time, either in revision or just reading over what you plan to do. You should not overdo revision at this stage, however, because there comes a time simply to be content you have it right and stick with it.

This whole preparation process is important and not to be skimped. Preparation does get easier, however. You will find that, with practice, you begin to produce material that needs less amendment and that both getting it down and any subsequent revision begin to take less time.

At the end of the day, as has been said, you need to find your own version of the procedures set out here. A systematic approach helps, but the intention is not to overengineer the process. What matters is that you are comfortable with your chosen approach, and that it works for you. If this is the case, then, provided it remains consciously designed to achieve what is necessary, it will become a habit. It will need less thinking about, yet still act to guarantee that you turn out something that you are content meets the needs – whatever they may be.

We shall now look at the presentation stage by stage and start, with appropriate logic, at the beginning, and see how you can get to grips with that.

6.6 The structure of a presentation

6.6.1 The beginning

The beginning is clearly an important stage. People are uncertain, they are saying to themselves, 'What will this be like? Will I find it interesting/helpful?' They may also have their minds on other matters: 'What's going on back at the office, the job I left half finished? How will my assistant cope when I'm away, even for a few minutes?' This is particularly true when the people in the group do not know you, or do not know you well. They then have little or no previous experience of what to expect, and this will condition their thinking (it is also possible that previous experience will make them wary!). With people you know well, there is less of a problem, but the first moments of any speech are nevertheless always important.

The beginning is important not only to the participants, but also to the presenter. Nothing settles the nerves – and even the most experienced speakers usually have a few qualms before they start – better than making a good start. Remember, the beginning is, necessarily, the introduction; the main objective is therefore to set the scene, state the topic (and rationale for it) clearly, and begin to discuss the 'meat' of the content. In addition, you have to obtain the group's attention – they will never take the message on board if they are not concentrating and taking in what goes on – and create some sort of rapport both between you and them, and around the group themselves.

Let us take these aspects in turn.

6.6.1.1   Gaining attention

This is primarily achieved by your manner and by the start you make. You have to look the part; your manner has to say, 'This'll be interesting. This person knows what they're talking about.' A little has been said about such factors as appearance, standing up and so on. Suffice it to say here that, if your start appears hesitant, the wrong impression will be given and, at worst, everything thereafter will be more difficult. More important is what you say first and how it is said.

There are a number of types of opening, each presenting a range of opportunities for differing lead-ins. Here are some examples.

- **A question**: It can be rhetorical or otherwise, preferably something that people are likely to respond to positively: 'Would you welcome a better way to …?'
- **A quotation**: This might be humorous or make a point, which might be a classic, or novel phrase; or it might be something internal: 'At the last company meeting, the MD said …'
- **A story**: Again, this could be something that makes a point, relates to the situation or people, or draws on a common memory: 'We all remember the situation at the end of the last financial year when …'
- **A factual statement**: This could be striking, thought-provoking, challenging or surprising: 'Do you realize that this company receives 120 complaints every working day?' (The fact that this is also a question indicates that all these methods and more can be linked.)
- **A dramatic statement**: This might be a story with a startling end. Or a statement that surprises in some way. For instance, once, talking about direct-mail advertising, I started by asking the group to count, out loud and in unison, from one to ten. Between two and three I banged my fist down on the table saying, '*Stop!*' loudly. 'And that', I continued, 'is how long your direct mail has to catch people's attention – two and a half seconds!'
- **A historical fact**: A reference back to an event that is a common experience of the group might be in order here: 'In 2000, when company sales for what was then a new product were just …'

- **A curious opening**: Simply try a statement sufficiently odd for people to wait to find what on earth it is all about: 'Consider the aardvark, and how it shares a characteristic of some of our managers ...' (In case you want a link, it is thick-skinned.)
- **A checklist**: This is perhaps a good start when placing the 'shopping list' in mind early on is important: 'There are ten key stages to the process we want to discuss, first ...'

There must be more methods and combinations of methods that you can think of. Whatever you choose, this element of the session needs careful, and perhaps very precise, preparation.

6.6.1.2 Creating rapport

At the same time, you need to ensure that an appropriate group feeling is engendered. In terms of what you say (participation also has a role here), you may want to set a pattern of 'we' rather than 'you and us'. In other words, say 'We need to consider ...' and not 'You must ...' If this approach is followed, a more comfortable atmosphere is created. You may add – discreetly – a compliment or two ('As experienced people, you will ...'), though without overboasting. And, above all, *be enthusiastic*. It is said that the one good aspect of life that is infectious is enthusiasm. Use it.

At the same time, the opening stages need to make it absolutely clear what the objectives are, what will be dealt with and how it will benefit those present. It must also move us into the topic in a constructive way.

This opening stage is the first 'Tell 'em' from 'Tell 'em, tell 'em and tell 'em', and directs itself at the first two stages of the group's thinking process.

6.6.2 The middle

The middle is the core of the session. The objectives are clear:

- to put over the *detail* of the message;
- ensure *acceptance* of the message;
- maintain *attention* throughout the process.

One of the principles here is to take one point at a time; we shall do just that here.

6.6.2.1 Putting over the content

The main trick here is to adopt a structured approach. Make sure you are dealing with points in a logical sequence, e.g. working through a process in a chronological order. And use what is referred to in communications literature as *flagging* or *signposting*, taking us straight back to the three 'tell 'ems'. You cannot say things like this too much: 'There are three key points here: performance, method and cost; let's deal with them in turn. First, performance ...' It gives advance warning of what is coming. (This applies to both content and the nature of what is being said. Saying 'For example ...' is a simple form of signposting. It makes it clear what you are doing and makes the point also that you are not moving on to the next content point just yet.) Putting everything in context, and relating it to a planned sequence, keeps the message organized and improves understanding.

This technique and the clarity it helps produce give you the overall effect you want. People must obviously understand what you are talking about. There is no room for verbosity, for too much jargon, or for anything that clouds understanding. One pretty good measure of the presenter is when people afterwards feel that, perhaps for the first time, they really have come to understand clearly something that has just been explained.

You cannot refer to 'manual excavation devices': in presenting, a spade has to be called a spade. What is more, it has, as it were, to be an interesting spade if it is to be referred to at all and if attention is to be maintained.

6.6.2.2   Maintaining attention

Here again the principles are straightforward. Keep stressing to the audience the relevance of what is being discussed. For instance, do not just say that some matter will be a cost saving to the organization, but stress personal benefits: will it make something easier, quicker or more satisfying to do, perhaps?

Make sure that the presentation remains visually interesting by using visual aids and demonstrations wherever possible. Use descriptions, too, that incorporate stories, or anecdotes to make the message live. You cannot make a presentation live by formal content alone: you need an occasional anecdote, or something less formal. It is good if you are able both to proceed through your content and, at the same time, to remain seemingly flexible, apparently digressing and adding in something interesting, a point that exemplifies or makes something more interesting as you go. How do you do this? It is back to preparation.

Finally, continue to generate attention through your own interest and enthusiasm.

6.6.2.3   Obtaining acceptance

People will implement only what they have come to believe is good sense. It is not enough to have put the message over and for it to be understood – it has to be *believed*.

Here we must start by going back to *understanding*, and noting that nothing will be truly accepted unless this is achieved. Note that to some extent better understanding is helped by the following.

- **Using clear, precise language**: This is language that is familiar to those present, and which does not overuse jargon.
- **Making explanation clear**: Make no assumptions, use plenty of similes (you cannot say 'This is like …' too often), and with sufficient detail to get the point across. One danger here is that, in explaining points that you know well, you start to abbreviate, allowing your understanding to blind you as to how far back it is necessary to go with people for whom the message is new.
- **Demonstration**: This adds considerably to the chances of understanding. Demonstrations can be specific: talk about products, for instance, and it may be worth

showing one. In this case, the golden rule is (surprise, surprise!) preparation. Credibility is immediately at risk if something is mentioned and needs visualizing, yet cannot be. Help your audience's imagination and your message will go over better.

- **Visual aids**: These are a powerful tool to understanding. As the old saying has it, 'A picture is worth a thousand words.' Graphs are an excellent example. Many people instantly understand a point from a clear graph that might well elude them in a mass of figures. Visual aids are commented on at the end of this section.

Effectiveness is not, however, just a question of understanding. As has been said, acceptance is also vital. Acceptance is helped by factors already mentioned (telling people how something will benefit them – or others they are concerned about, such as their staff), and the more specific this link can be made the better the effect will be on the view formed.

In addition, acceptance may come only once credibility has been established, and this, in turn, may demand something other than your saying, in effect, 'This is right.' Credibility can be improved by such things as references and things other people say. A description that shows how well an idea or system has worked in another department, and sets this out chapter and verse, may be a powerful argument. Always with references, this is dependent on the source of the reference being respected. If the other department is regarded in a negative way, then their adopting some process or product may be regarded by others as being a very good reason *not* to have anything to do with it. References work best when the results of what is being quoted are included, so that the message says they did this and so and so has occurred since, with sufficient details to make it interesting and credible.

Finally, it is worth making the point that you will not always know whether acceptance of a point has been achieved, unless you check. People cannot be expected to nod or speak out at every point, yet knowing that you have achieved acceptance may be important as you proceed. Questions to establish appropriate feedback are therefore a necessary part of this process, and in some presentations this must be done as you progress. It is also advisable to keep an eye on the visible signs, watching, for instance, for puzzled looks.

6.6.2.4 Handling objections

The first aspect here is the anticipation, indeed the pre-emption, of objections. On occasions it is clear that some subject to be dealt with is likely, even guaranteed, to produce a negative reaction. If there is a clear answer, then it can be built into the presentation, avoiding any waste of time. It may be as simple as a comment such as, 'Of course, this needs time, always a scarce resource, but, once set-up is done, time will be saved – frequently.' It would then go on to explain how this will happen.

Otherwise, if objections are voiced – and of course on occasion they will be – then a systematic procedure is necessary if they are to be dealt with smoothly.

First, give it a moment: too glib an answer may be mistrusted or make the questioner feel – or look – silly. So, pause …and for long enough to give yourself time to think (which you might just need!), and give the impression of consideration. An acknowledgement reinforces this – 'That's a good point'; 'We must certainly think about that' – though be careful of letting such a comment become a reflex and being seen as such. Then you can answer, with either a concentration on the individual's point and perspective, or a general emphasis, which is more useful to the group as a whole; or both, in turn.

Very importantly, never, ever bluff. If you do not know the answer you must say so (no group expects you to be infallible), though you may well have to find out the answer later and report back. Alternatively, does anyone else know? Similarly, even when you *can* answer, there is no harm in delaying a reply: 'That's a good point. Perhaps I can pick it up in context when we deal with ...'

A final word here: beware of digression. It is good to answer any ancillary points that come up, but you can stray too far. Part of the presenter's job is that of chairperson; everything planned for the session has to be covered, and before the scheduled finishing time. If, therefore, you have to draw a close to a line of enquiry, and you may well have to do so, make it clear that time is pressing. Do not ever let anyone feel it was a silly point to raise.

After all this, when we have been through the session, the time comes to close.

6.6.3   The end

Always end on a high note. The group expect it, if only subconsciously. It is an opportunity to build on past success during the session or, occasionally, to make amends for anything that has been less successful.

That apart, the end is a pulling together of the overall message that has been given. However you finally end, there is often a need to summarize in an orderly fashion. This may well be linked to an action plan for the future, so that, in wrapping up, what has been said is reviewed – completing the 'Tell 'ems' – and a commitment is sought as to what should happen next. This is important. Most people are under pressure for time and, whatever else, you have already taken up some of that. They will be busier after even half an hour taken to sit through your presentation than would be the case if they had not attended, so there is a real temptation to put everything on one side and get back to work – get back to normal. Yet this may be just where a little time needs to be put in to start to make some changes. Their having a real intention in mind as they leave the session is not a guarantee that action will flow, but it is a start. It makes it that much more likely that something will happen, especially if follow-up action is taken to remind and see the matter through.

As with the beginning, there is then a need to find a way of handling, in this case, the final signing-off. Here are some examples of ways to finish.

- **A question**: This could leave the final message hanging in the air, or makes it more likely that people will go on thinking about the issues a little longer: 'I asked a question at the start of the session; now let's finish with another ...'
- **A quotation**: This could encapsulate an important, or the last, point: 'Good communication is as stimulating as black coffee, and just as hard to sleep after' (Anne Morrow Lindbergh, American aviator, author).
- **A story**: This would be longer than the quotation, of course, but with the same sort of intention. If it is meant to amuse, be sure it does; you have no further

chance at the end to retrieve the situation. That said, I will resist the temptation to give an example, though a story close does not imply only a humorous story.

- **An alternative**: This may be as simple as 'Will you do this or not?' or the more complicated option of a spelt-out plan A, B or C?
- **Immediate gain**: This is an injunction to act linked to an advantage of doing so now. 'Put this new system in place and you'll be saving time and money tomorrow.' More fiercely phrased, it is called a fear-based end: 'Unless you ensure this system is running you will not ...' Although there is sometimes a place for the latter, the positive route is usually better.
- **Just a good close**: Alternatively, choose something that, while not linked inextricably to the topic, just makes a good closing line; for example, 'The more I practise, the more good luck I seem to have' (which is attributed to just about every famous golfer there is), is one that might suit something with a training or instructional content.

However you decide to wrap things up, the end should be a logical conclusion, rather than something separate added to the end.

All of this is largely common to any presentation. The importance of different presentations varies, however. Some have more complex objectives than others. In simple terms, you may want to inform, motivate, persuade, change attitudes, demonstrate, prompt action and more. Sometimes several of these are key together.

Consider an example. You want people to understand and take on board doing something differently. You want people not just to say that they understood the presentation and perhaps even enjoyed it: you want them to have learnt from it. The ways in which people learn are therefore important principles to keep in mind throughout. It needs patience as well as intellectual weight or 'clout'. It needs sensitivity to the feedback as well as the ability to come through it. As with many skills, the difficulty is less with the individual elements, most of which are straightforward and commonsense, than with the orchestration of the whole process. Many people in business must be able to present effectively, to remain flexible throughout and work with an audience rather than just talk at them.

Whatever it is you do, you *make it happen*. Thus you must *plan* to make it happen. You can rarely, if ever, 'just wing it': it needs care in preparation and in execution. If it is given appropriate consideration, you can make it go well. There are several other things to bear in mind, however, that will help.

6.7 Speaker's notes

For most people having *something* in front of them as they speak is essential. The question is what form exactly should it take? Speaker's notes have several roles. They:

- *boost confidence* – in the event you may not need everything that is in front of you, but knowing it is there is, in itself, useful;
- *act as a guide* to what you will say and in what order;

- *assist you* to say it in the best possible way, producing the right variety, pace, emphasis, etc. as you go along.

On the other hand, it must not act as a straitjacket and stifle all possibility of flexibility. After all, what happens if your audience's interest suggests a digression or the need for more detail before proceeding? Or the reverse: if a greater level of prior information or experience becomes apparent, meaning that you want to recast or abbreviate something you plan to say? Or what if, as you get up to speak for half an hour, the person in the chair whispers, 'Can you keep it to twenty minutes? We're running a bit behind'? Good notes should assist with these and other scenarios as well.

Again there is no intention here either to be comprehensive or to suggest that only one way makes sense. Rather, I will set out what seem to me some rules and some tried and tested approaches. But the intention is not to suggest that you follow what is here slavishly. Again, it is important to find what suits you, so you may want to try some of the approaches mentioned, but to amend or tailor them to suit your kind of presentation as exactly as possible.

One point is worth making at the outset: there is advantage in adopting (if not immediately) a consistent approach to how you work here. This can act to make preparation more certain, and you are more likely also to become quicker and quicker at getting your preparation done if you do so.

6.7.1   The format of notes

The following might be adopted as rules.

- *Legibility* is essential (you must use a sufficiently large typeface, or writing, avoid adding tiny, untidy embellishments and remember that notes must be suitable to be used standing up and therefore at a greater distance from your eyes than if you sat to read them).
- The materials must be *well chosen* – for you. Some people favour small cards, others larger sheets. A standard A4 ring binder works well (one with a pocket at the front may be useful for ancillary items you may want with you). Whatever you choose, make sure it *lies flat*. It is certain to be disconcerting if a folded page turns back on itself – especially if you repeat a whole section. It can happen!
- Using *only one side of the paper* allows space for amendment and addition if necessary and/or makes the total package easier to follow (some people like notes arranged with slides reproduced alongside to produce a comprehensive double-page spread).
- Always *page-number your material* (yes, one day, as sure as the sun rises in the morning, you will drop it). Some people like to number the pages in reverse order – 10, 9, 8, etc. – which gives some guidance regarding time remaining until the end. Decide which – and stick with one way to avoid confusion.
- *Separate different types* of note: for example *what you intend to say* and *how* (emphasis etc.).
- Use *colour and symbols* to help you find your way, yet minimize what must be noted.

Never put down verbatim what you want to say. Reading something is difficult and always sounds less than 'fresh'. The detail on the speaker's notes needs to be just sufficient for a well-prepared speaker to be able to work from it and do so comfortably. Consider the devices mentioned here, and try to bear in mind as you do so the effect that the use of a second (or third?) colour – which cannot be reproduced here – would have on its ease of use. Some highlighting is clearly more dramatic in fluorescent yellow, for example.

Consider these ideas; there should be things here you can copy or adapt, or that prompt additional ideas that suit you.

- **Main divisions**: The pages – imagine they are A4 – are divided (a coloured line is best) into smaller segments, each creating a manageable area on which the eye can focus with ease; this helps ensure that you do not lose your place (effectively, it produces something of the effect of using cards rather than sheets).
- **Symbols**: These save space and visually jump off the page, making sure you do not miss them. It is best to avoid possible confusion by always using the same symbol to represent the same thing – and maybe also to restrict the overall number used, since a plethora of them might become difficult to follow. Use bold explanation marks, for example, or 'S1' (Slide 1) to show which slide is shown where.
- **Columns**: These separate different elements of the notes. Clearly there are various options here in terms of numbers of columns and what goes where.
- **Space**: Turning over takes only a second (often you can end a page where a slight pause is necessary anyway), but it is always best to give yourself plenty of space, not least to facilitate amendments and, of course, to allow individual elements to stand out.
- **Emphasis**: This must be as clear as content; again, a second colour helps.
- **Timing**: An indication of time elapsed (or still to go) can be included as little or often as you find useful; remember that the audience love to have time commitments kept.
- **Options**: This term is used to describe points included as a separate element and such can be particularly useful. Options can be added or omitted depending on such factors as time and feedback. They help fine-tune the final delivery – and are good for confidence, also. They might go in a third right-hand column.

Note that points in the 'Options' column are designed to be included or not, depending on the situation. A plan might thus include ten points under options with half of them (regardless of which) making your total presentation up to the planned duration. Thus, you can extend or decrease to order and fluently work in additional material where more detail (or an aside, or example) seems appropriate on the day.

Good preparation and good notes go together. If you are well prepared, confident in your material and confident also that you have a really clear guide in front of you, then you are well on the way to making a good presentation.

A suggestion of the sort of planning sheet that might be helpful is in Panel 6.1. This is designed to act as both a checklist and a way of setting out your first thoughts.

Panel 6.1   Presentation planner

TOPIC or *TITLE*

Duration (specified or estimated):

My intentions are to:

My overall objective can be summed up as:

Summary of main points to be made:

STRUCTURE

The logic and sequence used will be:

The beginning:

Things to make clear:

Content:

The middle:

The end:

Final 'sign-off':

Additional points:

Set out headings like these, with space to make notes, and you are well on the way to planning a good presentation.

6.8   Visual aids

6.8.1   *The most important visual aid*

Perhaps the most important visual aid has already been mentioned: it is you. Numbers of factors, such as simple gestures (for example, a hand pointing), and more dramatic ones such as banging a fist on the table, which may be described as flourishes, are part of this, as is your general manner and appearance.

More tangible forms of visual aid are also important. Such things as slides serve several roles; these include:

- focusing attention within the group;
- helping change pace, add variety, etc;

- giving a literally visual aspect to something;
- acting as signposts to where within the structure the presentation has reached.

They also help the presenter, providing reminders over and above speaker's notes on what comes next.

Be careful. Visual aids should *support* the message, not lead or take it over. Just because slides exist or are easy to originate, it does not mean they will be right. You need to start by looking at the message, at what you are trying to do, and see what will help put it over and have an additive effect. They may make a point that is difficult or impossible to describe, in the way a graph might make an instant point that would be lost in a mass of figures. Or you may have a particular reason to use them: to help get a large amount of information over more quickly, perhaps.

The checklist that follows deals, briefly, with the various options, offers general guidance on visuals production, and some tips on using the ubiquitous OHP (overhead projector) and PowerPoint.

6.8.2   General principles of using visual aids

- Keep the content *simple*.
- *Restrict* the amount of information and the number of words: use single words to give structure, headings, or short statements; avoid it looking cluttered or complicated; use a running logo (e.g. the main heading/ topic on each slide).
- Use *diagrams*, *graphs*, etc. where possible rather than too many figures; and never read figures aloud without visual support.
- Build in *variety* within the overall theme, e.g. with colour or variations of the form of aid used.
- Emphasize the *theme and structure*, e.g. regularly using a single aid to recap the agenda or objectives.
- Ensure the content of the visual *matches the words* spoken.
- Make *necessary and relevant*.
- Ensure *everything is visible*, asking yourself, 'Is it clear? Will it work in the room? Does it suit the equipment?' (Colours, and the right-sized typeface help here.)
- Ensure the layout *emphasizes the meaning* you want (and not some minor detail).
- Pick the *right aid* for the right purpose.

6.8.3   Using an overhead projector

Given how the computer has usurped the overhead projector, or OHP, space will not be taken up here with advice; suffice to say they need to be used carefully and, if you have to use one, check it out – and practise.

The ubiquitous Microsoft PowerPoint system (and, to be fair, other systems) allows you to prepare slides on your computer and project them through a projector using the computer to control the show. So far, so good. It works well and you have the ability to use a variety of layouts, colours, illustrations and so on at the touch of a button.

There are some dangers (and many of the points made about using visual aids apply equally here). First, do not let the technology carry you away. Not everything it will do is useful – certainly not all on one slide or even in one presentation – and it is a common error to allow the ease of preparation to increase the amount on a slide beyond the point where it becomes cluttered and difficult to follow. This might also lead you to use too many slides. Similarly, if you are going to use its various features, such as the ability to strip in one line and then another to make up a full picture, remember to keep it manageable. Details here can be important: for instance, colour choice is prodigious but not all are equally suitable for making things clear.

The second danger is simply the increased risk of technological complexity. Sometimes it is a simple error. Recently I saw an important presentation have to proceed without the planned slides because the projector (resident at the venue) could not be connected to the laptop computer (which had been brought to the venue) because the leads were incompatible. Sometimes, problems may be caused by something buried in the software. Again, not long ago I sat through a presentation that used 20–30 slides and, each time the slide was changed, there was an unplanned delay of three or four seconds. It was felt unwarranted to stop and risk tinkering with the equipment, but, long before the 45-minute presentation finished, everyone in the group found it disproportionately maddening.

Always make sure (check, check, check!) that everything is going to work. Run off transparencies that can be shown on an OHP in the event that disaster strikes, if this would be a sensible insurance (or prepare a paper handout copy). Finally, follow all the overall rules and do not forget that you do not have to have a slide on all the time – when you have finished with one, blank out the screen until you are ready for the next.

Whatever you use, remember to talk to the group, not to the visual aid. Looking at the screen too much when slides are used is a common fault. Make sure visuals are visible (do not get in the way yourself), explain them or their purpose as necessary, mention whether or not people will get a paper copy of them and stop them distracting your audience by removing them as soon as you are finished with them.

6.8.4   Beware gremlins

Is this one of Murphy's Laws? Certainly, it is an accurate maxim that if something can go wrong it will; and nowhere is this more true than with equipment.

The moral: check, check and check again. Everything – from the spare OHP bulb (do not even *think* about using an old machine with only one bulb), to which way up the 35 mm slides are going to be, even to whether the pens for the flipchart still work – is worth checking.

Always double-check anything with which you are unfamiliar, especially if, as with a microphone for instance, what you do is going to be significantly dependent on it. And remember that, while the sophistication of equipment increases all the time, so too do the number of things that can potentially go wrong.

The concept of contingency is worth a thought: what do you do if disaster does strike? You have been warned.

6.8.5 *Anything and everything*

Finally, be inventive. Practically anything can act as a visual aid, from another person (carefully briefed to play their part) to an exhibit of some sort. In a business presentation, exhibits may be obvious items – products, samples, posters, etc. – or may be something totally unexpected. Engineers may well have things to show, ranging from mere plans to an actual item that is a vital part of their presentation.

Like all the skills involved in making presentations, while the basics give you a sound foundation, the process is something that can benefit from a little imagination.

6.9 Summary

There is a good deal of detail here, albeit mostly common sense. It represents a lot to keep in mind at one time. Practice and building up the right habits help. Overall, the key issues are:

- *preparation* in all its manifestations – this is simply a must;
- balancing *content* and *manner* – it is as much *how* you say things as *what* you say that determines the level of impact;
- taking *time* – you must allow yourself the opportunity to use techniques, not simply rush through the content to get it over;
- *visual aids* – they can help (indeed, may be expected), but they must support what is said rather than lead.

Chapter 7
Negotiation: making the best deal

I saw Annabel (aged six) quoted on the Internet: 'If you want a hamster, you start by asking for a pony.' This sums up superbly the concept of negotiating, and in this chapter we focus specifically on this key communication. The following defines the process and reviews both strategy and tactics, showing how to plan and conduct successful negotiations and secure the deal you want

Negotiation, the process of making a deal and agreeing the terms on which it is arranged, is an important, and ubiquitous, business communications skill. Annabel, quoted above, says something that sums up much about the negotiation process. Perhaps she has some inherent knowledge or insight into the process that may stand her in good stead in later life, because negotiation – bargaining, to put it simply – is used in so many different contexts in both private and organizational life.

It is essentially *an interactive communication skill*, one that must be deployed in many different circumstances and at every level of organizational life. It is a close partner of persuasive skills; it is as much used on the other side of the sales process, in purchasing, and may need to be utilized in a wide variety of business dealings, from union negotiation to corporate takeover and merger arrangements. Negotiation skills are not only necessary to deploy in doing many jobs successfully, but are also among those needed if someone is to be seen as fully competent; excelling in these areas enhances the likelihood of career success. Negotiation may also be needed to obtain the best deal – a remuneration package, for instance – for yourself.

Good negotiators are in a strong position to make a good impression and a good deal.

7.1 A means to an end

A great deal can be riding on the outcome of a negotiation. Success can make money, save time or secure your future (and your reputation). To negotiate and do so success-fully is to deploy a technique that can work positively for you in a host of different ways. The overall deal you strike may be vital, and individual elements of it can be significant, perhaps very significant. For example, this book could not have been written without agreeing a deadline that made it possible.

The techniques of negotiation are many and varied. It needs the right approach, the right attitude and attention to a multitude of details on the way through. Like so many business skills, it cannot be applied by rote: its use must be tailored – intelligently tailored – to the individual circumstances on a case-by-case basis. It has elements of being an adversarial process about it and it needs handling with care; individual techniques may be common sense in some ways, but they need deploying with some sensitivity. You can as easily find that someone is running rings around you as that you are tying up the deal of a lifetime.

In this chapter, negotiation is explained and investigated. The way it works is spelt out and the process is illustrated in a variety of ways and in different contexts. It shows that it is important not only to be able to negotiate, but also to be able to plan and manage the process in order to increase the likelihood of achieving the outcome you want.

7.2   A changing world

It should be said early on that negotiation is a frontline skill. It puts those undertaking it in an exposed position. It may involve people within the same organization or outside it; much negotiating is traditionally between supplier and buyer. Whatever the precise purpose of negotiation, it is affected by the increasingly dynamic and competitive world in which organizations operate. For example, buyers negotiating arrangements with suppliers have considerable power and there is always a competitor waiting in the wings to pick up the pieces if a supplier fails to make a deal that is acceptable. Such competitive pressure also exists internally within an organization and can affect all sorts of negotiation.

However much it may be a skill that needs to be deployed widely around an organization, and one therefore that many people should aim to have as a technique in their armoury, it is certainly often a high-level one. Senior management and leaders of many kinds must be good negotiators. If the skill is one that 'goes with the territory' for you, then there is real danger in failing to get to grips with it; and real opportunity for those who make it their stock-in-trade.

Negotiation is a complex process. At least it is complex in the sense that it is an interactive process that involves a multitude of techniques. Perhaps the greatest complexity is that of orchestrating the overall process, managing it within the context of a meeting. The purpose of negotiation is, however, clear.

7.3   A special form of communication

Negotiation is a special form of communication and is best defined in that context. Communication is the basic process, the flow of information between people that informs, instructs and more. More important here is *persuasive communication*. This is designed to produce agreement and action in another person; as such, this may have a wealth of applications, including selling where the agreement is to buy something (see Chapter 4).

When persuasion has worked and agreement is there, at least in principle, negotiation may take over. It is concerned with the relationship between two parties where the needs of both are largely in balance. It is the process of *bargaining* that arranges and agrees the basis on which agreement will be concluded – the terms and conditions under which the deal will be struck.

Consider two simple examples. In the classic case of wage bargaining, the employer wants to reach an agreement (to secure the workforce and keep the business running), and the employees want an agreement (so that the process of negotiating is over, and they can get on with earning at a new, improved, rate). This question of balance defines the process. In selling, the first stage is to get agreement – from the point of view of the seller *to get what they want* – but, beyond that, negotiation is what *decides the 'best deal'*. Thus, if you are buying a car, say, then the things that need arrangement are all those making up the 'package', which goes beyond just the car itself. Such factors may include: the finance, discounts, extras to be included with the car (air-conditioning, perhaps) that are not standard, delivery, trade-in of an existing vehicle. The conversation will revolve around a possible point of agreement, from a starting point – *initial stance* – to the ultimate *point of balance*. To be clear: the initial stance is the starting point or first offer in a negotiation, one that is almost always pitched high, sometimes clearly unreasonably high; the point of balance is the point where the 'deal' can be agreed by both parties, though it may not reflect the best hopes of either.

7.4   The application of negotiation

The applications of negotiation are wide; they include being used:

- as part of the *sales process* (by both buyer and seller); ditto persuasion;
- between individuals for *primarily personal reasons* (e.g. negotiating a pay increase or remuneration package or 'discussing' with your boss when you might go on holiday);
- in *wage bargaining* (as between an employer organization and a union or staff group);
- in *political circles* (as in treaties between governments);
- *internationally* (either between individuals or organizations in different countries or literally on a worldwide basis – like the recent talks about measures to combat global warming);
- in *corporate affairs* (takeovers, mergers and a variety of alliances and collaborations, sought or forced by circumstances).

It often involves a financial element (though it may not) and can involve two people or groups of people and take place at every level of an organizational hierarchy. Finally, it may be momentary and minor – 'If you can deputize for me at tomorrow's meeting, I can give you a little longer on that deadline we spoke about' – but still need getting right.

7.5   The nature of negotiation

Negotiation is characterized by various factors.

- It is an *interactive and balanced* process and one where the outcome must, by definition, be agreeable to both parties (though that does not mean both parties will necessarily regard the outcome as ideal); this is usually referred to as the *win–win factor*. In win–win negotiation the best negotiating outcome is one in which both parties are satisfied and, while they may not have what they would regard as the ideal 'best deal', they do have an agreement with which they can feel comfortable, and one that allows a good working relationship to continue if necessary.

- An *adversarial element* is inherent within the process, as each party vies to get the best deal that they can. Keep in mind sayings such as that of the author Ashleigh Brilliant: 'I always win. You always lose. What could be fairer than that?' This must be kept in check, though: if it gets out of hand negotiations may deteriorate into slanging matches with both parties making demands to which neither will ever agree, so that the whole process is stillborn
- A major part of the process of bargaining is one of *trading*, in other words as the terms and conditions are discussed – the *variables*, as they are called – they must be traded to create a balance on a basis of, 'If I agree to this, you will need to let me have that.' More of this later.
- A fair amount of *give and take* is necessary, and the to-and-fro discussion takes time; negotiation cannot be rushed (this is particularly so in some cultures, for example in the East).
- A *ritual element* is involved, i.e. negotiation must be seen to have done justice to the task it addresses; time is one element of this, as are a variety of procedural matters.

7.6   Three key factors

The process of negotiation involves juggling with three key factors: *information*, *time* and *power*. Let us consider these in turn.

7.6.1   Information

The old saying that information is power is certainly true in negotiation. Both parties want to know as much as possible about the other: the person (or people), their needs, priorities, intention and approach. A better understanding on one side or the other allows a more accurate deployment of the skills and gives that side an incontrovertible advantage.

7.6.2   Time

This is always a pressure, and urgency and specific deadlines may be imposed on any situation, often coming from outside the control of the person negotiating.

For example, someone's boss may be imposing tight timing (for reasons explained or unexplained). Similarly, circumstances may affect timing, for example in the way the publication of a company's annual accounts – announcing record results – might make concluding a pay deal ahead of the announcement a priority for management.

On the other hand, time and timing are some of the raw materials of most negotiations and it is said, with some truth, that there has not been a deadline in history that was not negotiable.

7.6.3 Power

Many factors can add weight – power – to the ability to negotiate. The phrase about 'having someone over a barrel' picks up this point: it means power is very much on one side. Power stems from two main areas.

1. **The power of precedent**: This is the equivalent of the 'self-fulfilling prophecy' – we know something cannot be done because a certain past experience tells us so. The result? We avoid even raising an issue and the power moves to the other side. Negotiation demands an open mind, a thorough search for everything that might assist – taking a chance or a risk is part of the process and doing so and addressing every possibility regardless of precedent gives us power and improves the chances of success.
2. **The power of legitimacy**: This is power projected by authority. People's attitude to what can be negotiated comes, in part, from where and how they see something originate. For example, even something as simple as a form or a notice is often taken as gospel. Checking into a hotel, how many people do other than fill in the complete registration form? Very few. Most take it as a given that it must be done, yet, in my experience, if you ask, often a few key details are sufficient. The point here is that the authority may be real or it may be assumed or implied. In other words power is intentionally invested in something to give it *more* power and make it weigh more heavily in the balance. This may be very minor. Someone says, 'That would be against policy,' and suddenly someone else feels less able to challenge it. Even when both parties understand this, the principle still adds an additional element to discussions that, all the time, must assess the real power being brought to the table as the meeting proceeds. Thus, a case has legitimacy if is supported by factual evidence; the better the evidence, the higher the legitimacy and the stronger the case. Additionally, power comes from the legitimacy of factual evidence.

7.7   A constructive process

It is now worth looking at things from the other side and considering what negotiation is not. It is not an argument. A complaint makes a good example. Say your fridge is acting like a microwave oven and you go into the shop that supplied it to state your grievance. That is not negotiating, either. It may produce an apology, but what you want is action. There are numbers of things that could be done (ranging from swapping it for a new machine today, no questions asked, to getting in touch with the manufacturer, the latter implying a delay of indeterminate length). The mix of action, timing, recompense, etc. needs negotiating.

So negotiation demands that proposals be made and discussed. In the simple example above, negotiation can fail not because it is done badly but because it is avoided and not done at all. The situation is then likely to turn acrimonious and argument is all that ensues.

At the end of the day all the parties to negotiation need to understand its nature. It may be adversarial, but we are aiming at a mutually agreeable outcome – what has already been referred to as a *win–win* outcome. If one party goes headlong for the best deal regardless, the likelihood is that they will push the other into a corner and that they will feel unable to agree to anything; the negotiation stalls. If both parties accept that some compromise is necessary, then the outcome is likely to be better for both.

Thus negotiation is about seeking common ground, relating to the other person and their concerns, participating in to-and-fro debate, but not insisting on a totally rigid agenda. It means asking questions – and *listening* to the answers, disclosing information (to some extent), openly stating a point of view, building a relationship and treating the other person with respect.

Negotiation must aim throughout at agreement and actively act to avoid stalemate. If persuasion answers the question, 'Will this person agree?' then negotiation must address the question, 'On *what basis* will this person agree?'

Above all, understanding and utilizing negotiation require:

- a basis of sound, effective *communication skills* (because negotiation is a specialist form of communication);
- an understanding of the *role of negotiation* (because it is almost always part of a broader picture, for instance one that starts with persuasion);
- the ability to orchestrate a *plethora of techniques* and relate what is done to the particular meeting and circumstances (in other words this is not something that can be applied by rote);
- a *sensitivity* to the people involved, as what is done must be based on an understanding of them and their needs.

Already we see that negotiation, a specialist form of personal communication, and one that often goes hand in hand with selling, can secure a deal, obtain the best financial outcome and provide the basis of good business and personal relationships; but it must be done right.

Next, we set out an overview of how negotiation works, and review this in two main sections: first looking at the process, then at the tactics it involves.

7.8   The process of making the right deal

Let us be clear: negotiation and persuasion are different things. They are certainly interrelated: successful persuasion gains agreement to action (to agree a purchase, perhaps); negotiation is concerned with identifying, arranging and agreeing the terms and conditions that accompany agreement. Thus, persuasion and agreement must logically come first. People do not waste time negotiating about something that they have no interest in. That is not to say that prior agreement is always openly stated. It may well not be, and in this way persuasive communication and negotiation merge, with an imprecise line between them.

Negotiation is a very particular process, characterized in a number of ways.

- First, it is complex. The complexity comes from the need to orchestrate a many-faceted process rather than because of anything individually intellectually taxing. But you need to be quick on your feet as it were to keep all the necessary balls in the air, and always see the broad picture while concentrating on individual details.
- Second, while negotiation is not to be treated as an argument (if it is, then an impasse usually results), it *is* adversarial. Both parties involved want the best deal they can obtain. Yet compromise is essential: stick out for a perfect deal and the other party may walk away; give way too easily and you may well regret what is then agreed. Both must be satisfied with what they do agree (the win–win option).
- Third, there is a ritual aspect to negotiation. It is a process that needs to be gone through. It takes time. There is to-and-fro debate, and a mutually agreeable solution needs to be seen to be being sought out, as well as actually taking place. Too much haste, a rush for agreement or a take-it-or-leave-it approach can fail less because the deal it offers is unacceptable, more because the other party does not feel that the process is being taken seriously. People look for hidden meaning, believe that something better must be possible and, again, the outcome can be stalemate.

Because of these factors the best negotiators are at pains to take the broad view, to understand the other person's point of view and what they are trying to achieve and why. Because the process of negotiation deals with a complex mix of issues and motivations, the way this is handled, not least the confidence with which it is seen to be handled, is important. The negotiator who seems confident, and has an ability to deal with all the issues logically and to manage the overall process as well as picking up the detail, commands respect. How do you get on top of it all to this extent? Well, apart from having a clear understanding of the process, the key is preparation. It stands repeating: successful negotiating does not just happen; it is rarely possible to 'wing it'. Negotiation is not something most of us can make up as we go along. Remember the well-known saying of Abraham Lincoln: 'If I had nine hours to cut down a tree, I would spend six of them sharpening my axe.'

7.9   First things first

Successful negotiation begins with preparation. The rule here is simple: do it! Preparation may only be a grand term for the age-old advice that it is best to engage the brain before the mouth, and it may take only a few minutes. Of course, at the other end of the scale it may mean sitting round the table with a few colleagues thrashing out exactly how to proceed with something. Whatever the necessary scale, the rule is that it should always happen.

It is particularly important to have clear *objectives*. That is, having a clear specification – the *result* you want from the negotiation spelt out: what it is, its timing, etc.

If it is said simply that 'We want the best deal possible,' this provides nothing tangible with which to work. There is all the difference in the world between an author saying, 'Let's see if the publisher will pay me more for my next piece of text' and setting out 'to obtain a 10 per cent increase in the fee'. Planning should be designed to produce the equivalent of a route map, something that helps shape the meeting. You know you cannot anticipate everything. With people it is just not possible to predict exactly what will happen. Your plan should, however, provide both an ideal route forward and a basis that will assist you if things do not go exactly to plan.

A final point here may also encourage you to spend a moment in preparation. You need not only to be prepared, but also to *appear* well prepared. If it is obvious you are not, if it seems you are unfamiliar with the issues – and more so if this is so – then it is more likely someone will run rings round you.

Preparation is the foundation to success and insurance against being outclassed.

Bearing all this in mind, it is logical that the choice of who will undertake negotiation in particular circumstances can also be important, just as it makes a difference who will do best given a sales role, for instance.

7.10   The core element

The core of the negotiation process revolves around what are called *variables*: those factors that can be varied and arranged in different ways to create different potential deals. Thus, in negotiating price, for example, the price itself is clearly a variable, but discussion may involve associated matters such as payment terms, extras (e.g. with a product such might range from delivery to service arrangements to credit) and other factors such as timing and staffing; and more.

The overall rules here include:

- aiming high – going for the best deal possible;
- discovering the full list of variables the other person has in mind;
- treating *everything* as a potential variable;
- dealing with detail within the total picture (rather than one point at a time without reference to others).

Your use of the variables must increase the power from which you deal. You can use them in various ways. You can prompt attention by offering reward: something you are prepared to give. Conversely you can offer punishment: by flagging your intention to withhold something. Your case is strengthened, given power – or legitimacy in the jargon, by being linked to authority or supported by factual evidence. Similarly the use of *bogeys*: that is red herrings, elements that are apparently a significant part of the negotiation, which are introduced only to distract or confuse the issue. They are peripheral factors included only to distract or seek sympathy (e.g. statements such as

'That's beyond our control') aiming to stop questioning in its tracks regardless of its truth, and thus acting to add power.

You have to rank the variables, both in preparation and during the negotiation, when realistically some fine tuning may be necessary. There will be, at the very least, some things that are:

- *essential*: you cannot agree any deal without these points being part of it;
- *ideal*: what you intend to achieve (and the priorities, because there may be more of these than it is realistic to achieve);
- *tradable*: in other words, those things that you are prepared to give away to help create a workable deal.

The concept of *trading variables*, the process of deciding how factors are decided and agreed in relationship to each party to the negotiation, is key to the whole process of negotiation. It is important never to give anything away.

Concessions, a variable that is offered to the other party (usually in a way that balances the total picture) as part of the to-and-fro process of agreeing a total arrangement of variables, must be traded. For example an external consultant might say, 'We can certainly make sure all rail travel cost is second-class, but we do need to add a little to the fees for the time taken.'

Overall, the variables are exactly that: variable. Imagine a balance – negotiating parties are weighing up the deal they will, or hope to, get – with each variable represented by a block shape. As the negotiation proceeds, the shape of each, and thus the overall balance, changes until each side represents what someone can accept, with one probably a better deal than the other.

In this trading, the value of every concession must be maximized when you give it, and minimized when you accept it, so that the trading drives the balance in the direction you want. Thus, saying, 'I suppose I could do that, though it'll make more work, but OK' makes it seem that what you are agreeing is worth more than perhaps it actually is. However, saying, 'I would never normally do this' implies that you are making an exception in their favour. Similarly, saying, 'Well, I suppose if I do that you won't need to …' amplifies the effect that the concession seems to have for them.

Clearly, the way such things are said, perhaps incorporating a degree of exaggeration, in turn affects their reception.

It is similar with how you minimize the other parties' concessions. These can be:

- dismissed – 'Fine, now next …';
- belittled – 'Well, that's a small point out of the way';
- amortized – 'I suppose that saves a little each month';
- taken for granted – 'I would certainly expect that.'

In fact, they could be treated in any way to reduce them in power by the way they are accepted and referred to during the discussion.

So, the discussion has to be planned, directed and controlled. The confidence displayed during it is a significant factor (and links back to preparation). You must be clear about what you want to achieve. If you utilize every possible aspect of the discussion and treat everything as a variable, and deploy appropriate techniques to balance the whole picture and arrive at where you want to be (or somewhere close), then you will be able to achieve a reasonable outcome. Remember the win–win scenario. The job is not to take people to the cleaners. If you are prepared to agree only something that is weighted heavily in your favour, then negotiation may break down and no agreement at all may be concluded. Sometimes you need to be prepared for this. There is often a minimum arrangement below which you are unprepared to go, and sometimes walking away rather than agreeing something you are not prepared to live with is the right decision.

Sometimes, even if you have someone over the proverbial barrel, a widely skewed deal is best avoided. You need to think long-term. What will screwing them into the ground make you look like? What are the future consequences of forced agreement? What may happen next time when you do not have quite so much strength to bring to the party?

7.11   Techniques to add power

Describing the process thus far has omitted one important aspect, and that is the individual techniques that can be brought to bear. A confident negotiator may use a number of ploys to enhance what they do. Some are simple – for example, the use of silence, which many find embarrassing, to make a point or prompt a comment. Too often someone will ask something such as, 'How important is this to you?' They wait a moment and then continue, 'Well, I'm sure it must be an important factor, now let's …' They produce no real impact and, more important, no information by so doing; indeed silence may be joined by embarrassment. Wait. You can wait a long time if necessary (count to yourself to try this – you may well find that the pause that worries you and makes you feel you must continue speaking if they will not is only a few seconds, but just seems long). But using – really using – silence is one particularly significant ploy in negotiating.

This kind of element – negotiating tricks of the trade if you will – can enhance the process, turning a routine discussion into one that moves purposefully towards achieving your objectives. It is to this we turn next.

7.12   The tactics of negotiating: key techniques

Next we review ways of bringing control and power to what you do in aiming to strike a deal.

That negotiation is a complex process – certainly in the sense that there are many things to work at together if you are to make it go well – is clear. It was noted earlier that at its core it is an adversarial process, though balance is important and the concept of a 'win–win' deal that both parties can walk away from with some satisfaction is important. Much of what must be done revolves around the trading of variables – a process of 'if you will do this, I will do that' – which must take place for an acceptable balance to be reached. Such is often prompted by *what-if questions*, that is a specific aspect of negotiation, the process where adjustments are made by making suggestions that offer new ways of rebalancing matters: 'What if ...I do this and you then accept/do that?' Complex negotiations involve a good deal of this.

On the way, techniques and the careful deployment of various behavioural factors (with everything assisted by judicious preparation) can make or break your success.

With clear objectives in mind and an overall plan, you can begin discussions. An agenda is sensible for any complex meeting. There is some merit in being the one who suggests one, albeit as something helpful to both parties. If you say something like, 'We might find it best to ...' followed by an outline of how you want things to run (though *not*, of course, stating it as helpful to you), this sets the scene, though it must be borne in mind that this has something of the 'laying all the cards on the table' feel to it, so you may want to judge its precision and comprehensiveness carefully to allow you some flexibility.

With discussions under way, a variety of techniques can be used. None may be a magic formula, but together they add substance to how you are able to work. Being prepared to hold a silence (as mentioned earlier) is a good example. Other such devices include:

- *keeping track*: never lose the thread or your grip on the cumulative details; it helps to summarize regularly and to make notes as the meeting progresses;
- *being seen to be reasonable*: you can keep the perception of your attitude and of progress to date positive simply by the tone of voice and phrases you adopt: 'That's a good point – good idea, let's do that, it should work well';
- *reading between the lines*: nothing may be quite what it seems – or sounds – and you need to watch for 'danger phrases' from the other side, e.g. 'That's everything' (meaning 'everything for now, but there are more demands to come'); 'Right, that's good all round' ('and especially good for me');
- *focusing on the arrangements*: if you want a deal, then you must proceed as if there were potentially one to be had; making it sound as if you may not proceed at all casts doubt on the whole process;
- *concentrating*: keep thinking and run the conversation to allow this, building in pauses if necessary, e.g. say, 'Let me think about that' and pause, make a telephone call – whatever. But never be rushed ahead on an ill-considered basis;
- *considering matters in the round throughout*: be careful not to go for final agreement only to find out that the other party is still introducing new issues; there is a particular danger in agreeing to parts of the proposition one by one and then finding you are left with nothing to bargain with towards the end;
- *always regarding timing as a variable*: deadlines, duration and every other aspect of time are almost always negotiable;
- *always questioning what is described as fixed*: what seem like constraints can often be made into variables and included in the trading.

Any such list (and it could well be extended) quickly makes the point about the juggling trick that negotiation usually presents. The principles are individually straightforward, but there are a fair number of them, so the orchestration of the whole process needs skill and benefits from practice. It is important, therefore, to keep control. Containing emotions if necessary, and certainly remaining neutral, organized, not being rushed and being prepared to question matters as they are introduced are vital.

Overall, you should aim to run the kind of meeting you want, and create one that the other party will see as businesslike and acceptable.

The list above makes clear both the complexities of negotiation and the opportunity presented by attention to detail. One element of technique overlooked can affect things drastically. Consider a buying/purchasing situation (particularly one of some complexity and high value, such as might take place as a construction firm sought to purchase machinery and parts). If either party is dilatory in taking notes, then two things could happen. First, they may miss some detail and, proceeding without it, make less good arrangements than might otherwise be the case. Or, second, if they check, they get the detail but the other person now knows something new about them – that they are not concentrating or managing the detail as well as they might – and this alone may give them added confidence and allow them to score points that they would not do otherwise.

It makes the point that a key word to apply to the negotiation process is *fragile*.

Details matter, of course, both ways, since one person's slip up is another's opportunity. The link here with preparation is again clear. Being well organized, having thought it through and having a handle on how you will orchestrate all the different and often disparate factors are crucial. And it is not only the details to be negotiated that must be borne in mind: so too must the process. In the 1970s, when then (Tory) Prime Minister Edward Heath was faced with a major miners' strike, he refused to negotiate, making a take-it-or-leave-it offer, which in turn created ill feeling and extended the strike. The union, not unreasonably, simply did not believe him – that is not how negotiation works – and conducted itself on the basis that it could *not* be the final offer. Result? Deadlock.

Negotiation works best between professionals, when both parties accept the need for the process and take it seriously.

Remember that any confidence and professionalism you project will position you as someone to be respected, and that in turn may prevent certain more outrageous demands even being voiced. For instance, refusing to get hung up on something (even though it is, for you, a key issue), but, rather, being prepared to bypass it – 'To avoid getting bogged down, let's leave that, it's not so important; if we discuss so and so next we can pick it up later' – can impress. And so too can remembering to slot it in later, picking it up at an appropriate point (though the other party may hope it has been forgotten) and dealing with it in a different way to avoid stalemate.

7.13   Interpersonal behaviour

The thoughts above – keeping control, with emotion under wraps and so on – bring us naturally to the behavioural factors involved. Again, there are many, but the following are selected to give a feel of how they affect matters.

- **Disguised motivation**: Consider spelling out your true meaning and asking others to do the same. It is possible to have so much double guessing going on – 'Why are they asking that?' – that no one knows what is happening.

- **Advance warning**: The above can be helped by what many refer to as clear 'signposting', that is the technique of flagging ahead something of what is to come, describing either the topic or the nature of the next comment. For instance, this latter might be as simple as starting something 'For example …', making it clear that it is one among several options. This is best done positively: 'It might help meet the timing considerations if …', so that it is clear why a suggestion is being made. Negative signs ahead of a counterargument are often just a cue for the other person to stop listening as they begin to marshal their response: 'I'm not letting them get away with that, I'll suggest …'
- **Tactics to disrupt**: It should be recognized that not only is all fair in love and war, but that negotiation can be regarded as coming into this category also. Some – many – things are done to throw you (and you may want to act similarly). Examples include throwing out a smokescreen of many demands with the intention of getting agreement to one key demand hidden among others that are, in fact, less important. Or the use of flattery or coercion, an outburst (contrived) of anger, disbelief or outrage, pretended misunderstanding, playing for time (or imposing an unrealistic deadline on discussions), and more.
- **Giving nothing away**: Meant here in the sense of poker-playing impassiveness. If you sound firm, you must look firm – even if you are wondering what on earth you can do next. There is a link here with body language, which, while something of an uncertain science, may be worth a little study.

On the other hand, rather than disrupting, some points need to be powerfully made and are in the nature of attack. If this is the case, then it is important not to allow people to be put on their guard. This may be easily done, perhaps out of sheer politeness, with a circumspect statement such as, 'Look, I'm sorry to insist but this really is something I must handle carefully.' There are things about which we must be much more direct: 'It is impossible for me to go that far.' Full stop.

7.14 Keeping on track

Negotiation is just a special form of communication. So all the rules of good communication apply if negotiation is to be successful, and some of them are key.

Taking the need for clarity as read (if not always faultlessly achieved!), perhaps two are key. They go together and are asking questions and listening.

7.14.1 Questioning

Successful negotiators never try to proceed on the basis of hunch or assumption. If something is said that might be ambiguous, then they check it. Whether it is better to do this head-on or obliquely is a question for individual circumstances, and both will have their place. But, one way or another, negotiation must proceed on the basis of clear understanding.

Ask. Use open questions (i.e. those that cannot be answered yes or no), as these are most often best at obtaining information quickly and precisely. Get people talking. Pursue their real meaning and feelings, if necessary with a series of linked questions, and build what you discover into the plan you have made, adjusting how you implement it and fine-tuning as you go.

7.14.2   Listening

This is equally important. The great danger is to allow your mind to wander, albeit constructively, as you think ahead, when in fact this is not useful unless you have first picked up every detail and every nuance of what is said. There is no harm in being thought of as someone who misses nothing, whether on or between the lines. People are more careful if they think they are negotiating with someone who is confident and professional. Listening seems obvious, yet needs working at, and because it makes a real difference some consideration of how you can make sure it is effective is worthwhile (see Chapter 6).

As the variables are juggled to and fro, remember that there will be an intention behind everything said and done. If, early on, someone repeatedly says that some element of the discussion is not negotiable, it could mean just that. Or it could mean they hope to persuade you not to seek to use it as a variable even in a small way, even though they would be prepared to do so if pushed. If so, then a good tactic is to leave it on one side, gain agreement on other issues and establish a rapport with them before going back to test just how strong the resolve about it is in the face of an attractive agreement developing.

If someone reacts with shock, horror and surprise to a suggestion, you may have genuinely taken them unawares. Just as likely they hope to agree a rapid concession on the back of their exaggerated response. If so, then ignoring the tenor of the first response, and asking for a more considered comment might provide a better basis on which to move forward.

Such ploys and responses are the nuts and bolts of negotiating. Their permutations are virtually endless. But, whatever form they take, they are better dealt with from the firm foundation of a considered understanding of the negotiation process and how it works. For some people in business, one aspect of the profile of an individual may well be how they are seen as negotiators. Again, this can take many forms. It might be the admin department, seemingly with no feel for commercial details, securing an especially keen deal with a supplier. It might be the salesperson boosting profitability with a keen deal. Or it might be an accountant who, impressing a client with the apparently effortless professionalism with which they strike a deal on their behalf and in their presence, raises their profile sufficiently to prompt the client to allow them to act on their behalf again on other issues. In many different circumstances negotiation can achieve a variety of different things – both corporate and personal.

Overall, the key things to remember are that you must:

- *cope with the complexity* – and this means having a sound understanding of how the process works;
- *manage the discussion* – and this means taking time to prepare and keeping a steady hand on the tiller as it were throughout the discussion;

- *focus on the other party throughout* – because everything about what you do needs to be based on as good an understanding of the other party's needs, style and tactics as you can discover (both before and during discussion).

7.15 The scope of negotiation

To negotiate successfully, you must see the process in the round, take a broad view and continue to do so throughout the process. This means you must have a good grasp of the principles involved, for it is that which allows you the opportunity to orchestrate and fine-tune the process as you proceed. Small adjustments as you progress can make all the difference.

Negotiation is a big topic, so here are two checklists designed to encapsulate the essentials that make the process work in practice so much as space permits.

7.15.1 Checklist 1: Summarizing the principles

1. Definition: negotiation is about bargaining to reach a mutually agreeable outcome. This is the win–win concept.
2. Never neglect your preparation. Have a clear plan but remain flexible.
3. Participants must regard each other as equals. Mutual respect is essential to both conduct and outcome.
4. There is a need to abide by the rules. Negotiation is about discussion, rather than debate. There is little place for overt one-upmanship or domination, yet each must fight their corner.
5. Put your cards on the table, at least on major issues. Do not pretend powers or state intentions that are untrue.
6. Patience is a key characteristic of the good negotiator. Take your time; do not rush discussion or decision-making. Delay is better than a poor outcome.
7. Empathy is vital. Put yourself in the others' shoes. See things objectively from their point of view.
8. State clear objectives. Being open early on about overall intentions can save groping in the dark.
9. Avoid confrontation. Do not get into a corner you cannot get out of. Avoid rows and showdowns, but stand firm and keep calm.
10. Treat disagreement carefully. Act as devil's advocate, apparently looking at the case from the other's viewpoint, to avoid a confrontational style prompting disagreement.
11. Deal with concessions progressively. Where concessions have to be made, make them unwillingly and one at a time – and trade them.
12. Do not let perfection be the enemy of the good. An outcome that is 100 per cent what you want is rarely an option. Be realistic; do not waste time and effort seeking something out of reach.
13. Use openness but not comprehensively. Declaring your plans and intentions may be useful to the discussion. You may want to keep hidden the motivation behind them.

14. Stick with your objectives. Set your sights high and settle as high as possible. Know when to drop the whole thing rather than agree a totally inappropriate deal.
15. Keep up your guard. Maintain your stamina, bide your time. The other party may persevere for hours to see when you will crack.
16. Remain professional. For example, respect confidences that are given in the course of negotiations. Such consideration builds relationships and may help you next time.
17. Never underestimate people. A velvet glove may be disguising an iron fist.
18. End positively. Neither party will get exactly what it wants, but, if the deal is agreeable, emphasize this at the end.

7.15.2   Checklist 2: Summarizing the tactics

Like any interactive skill, negotiating is dependent on a plethora of factors. The following are picked to provide a top ten of things likely to be most useful. You might like to compose your own list, see how it varies, and make sure it reflects exactly the kind of negotiating you do and the kind of people it pits you against.

1. Select the right starting point. Your plan should make it easy for you to take the initiative and quickly get onto your agenda.
2. Aim high, then the trading moves you less far from what you regard as a good position.
3. Do not make your feelings obvious. There is an element of bluff. If your face and body language say, 'This is minor' as you respond to something major you will do better.
4. Use silence. Some things demand no reaction at all.
5. Watch for early difficulty. Let a rapport and momentum build up before you tackle contentious issues.
6. Do not exaggerate facts. They can be verified, and exaggeration causes problems later.
7. Communicate clearly. Remember the need for understanding as a foundation to the whole process.
8. Be seen to go with the other person's way of doing things, at least to some degree, and particularly if you are on their ground.
9. Do not push too hard. There is usually a line beyond which the outcome is not a better deal, but complete breakdown.
10. When negotiation is finished, stop. Once agreement is reached, clear, agreed and perhaps noted, move on to other matters. Otherwise people say, 'I've been thinking ...' and you are back to square one.

The importance of different factors such as these depends on the nature of the negotiation. Something full of complex technical or financial details poses different problems from something simpler.

Finally, you should note a few things to avoid. You will only excel if you never:

1. *overreact* if responses are negative – the other person is at pains not to say how excellent every point is;

2. allow yourself to become *overemotional, unpleasant, provocative or insulting* – a planned and controlled display of emotion may be useful, but you must know what you are doing;
3. agree to *something you do not want* – in many situations there is a minimal deal, which your plan should identify, below which it is better to walk away.

Sometimes, one key factor influences things disproportionately. For example, a sponsorship deal was once being negotiated by a famous American sportsman with a sports-clothing manufacturer. His face would be used in promotion, he would make some appearances and they would pay him well during the period that the deal ran. He wanted more money than the company wanted to pay. They wanted an agreement fast to use the arrangement in a new product launch, already scheduled. By agreeing to the man's face being used in this one-off campaign, for a reasonable fee, while overall negotiations continued, the sportsman's agent put him in a very strong position: as the launch broke and his face appeared across the nation, the company realized that they had either to agree terms or to explain to the public why he was no longer 'the face of the product'. Smart agent? Or perhaps a lapse of concentration by the company people fronting the negotiation? No matter, the interaction of timing and money was instrumental in settling the deal.

This kind of situation is just as common in less high-profile business situations. Again, it emphasizes the need for preparation and keeping track of things as negotiation proceeds, and, above all, doing so in an organized way.

Every negotiating situation you face can teach you something: what works well, what should be avoided, what best fits your style. The detail is important. Sometimes what makes the difference between success and failure is small and seemingly insignificant. One phrase, even one gesture, may make such a difference. If all the details are right, the whole will be more likely to work well.

Negotiation, or rather well-handled negotiation, can be very useful. When push comes to shove, a considered and careful – indeed, watchful – approach, systematically applied, is probably best; and remember the saying attributed to Lord Hore-Belisha: 'When a man tells me he is going to put all his cards on the table, I always look up his sleeve.' This sentiment should be regarded as good advice by any good negotiator.

7.16 Summary

Overall, the key issues are these.

- Negotiation is the process of making a deal and agreeing the terms on which it is arranged.
- It involves juggling information, time and power.
- The primary aim is usually an agreement acceptable to both parties.
- The key process within negotiating is trading variables.
- The nature of the process means that the details are important, and sound preparation and careful deployment of the many techniques are what make it work.

Chapter 8
Telephone communication: its special nature

This chapter explores how telephone contact is, by its nature, different from other forms, such as face-to-face contact (when visual clues add to understanding). It reviews the specific, and detailed, techniques needed to create fluent and fluid dialogue over the telephone and the specific elements, such as switchboards, mobile phones and voicemail, involved.

8.1 The nature of voice-only communication

As well as prompting a smile, James Thurber's comment, 'Well, if I called the wrong number, why did you answer the phone?', makes an immediate point. Telephone communication is not like any other form.

The following is something you recognize instantly: 'Er, yes …Hold on …Sorry, will you repeat that? …Well, it's a bit difficult, y'know. Just a tick …' It is the abominable no-man of the telephone; it is the moment when, even after only a few words, you *know* the communication is going to be difficult.

Yet the reverse is true too. Some people, also within a few words, create confidence. They make you feel that all will be well, that a good start has been made and get you looking forward to what comes next.

The telephone is ubiquitous and has been around a long time. It now comes not only as a simple phone, but as part of a profusion of electronic wizardry that can store information, carry out umpteen functions at once and play you music if you are kept waiting. It may be digital, portable and life may be unimaginable without it, but it should not be taken for granted. Good telephone communication does not just happen. It needs thinking about. It needs planning, and communication by phone needs executing with care.

Why? Because it is a powerful medium. The telephone can create a strong, perhaps lasting, impression – and do so for good or ill. It can typecast the speaker and their organization as efficient, helpful, positive and more; or set them apart as an offputting sign of an uncaring corporate dinosaur. This is possible at all levels and in all ways: the switchboard, someone dealing with a colleague in another department, external contact with customers we want to impress or suppliers on whose goodwill we are dependent – all such communications are influenced by telephone manner and thus by telephone skill.

Hence this chapter: there are principles and techniques involved that, properly deployed, can make this a major opportunity area for improved communications and better image-building. If everyone in an organization has been well briefed, understands the difference that can be made and uses the techniques consciously to create an impact that is appropriate, then the benefits can be considerable. Good telephone technique can:

- create an appropriate and positive image;
- avoid communications breakdown, and thus delay, confusion and perhaps waste of time and money;
- smooth the whole communications process on which any organization is dependent, internally and externally.

Particularly it must achieve clarity and project the right image.

8.2　An inherent fragility

Not only is it important to get telephone contact right. Indeed, there is more to it than a simple resolve to speak carefully. The process is inherently fragile: that is, small differences along the way can make a major difference to the impact.

One example, which illustrates this point well, occurred while I waited in the reception area of a company, having arrived a few minutes early for a meeting. The receptionist also staffed the switchboard and it quickly became apparent that she was very busy. Calls came in seemingly every few seconds and were clearly mostly from customers. The company had two separate departments that were predominantly concerned with customer contact: a sales office, which took orders, and an order-processing department, which saw to the subsequent details.

The switchboard operator had to discover, and discover fast, which department to put people through to. She was therefore repeatedly asking callers, 'Are you placing an order or chasing an order?' It worked. At least it gave her the information she needed to direct the call through accurately to the right department.

But did the company *want* every customer who telephoned to have it made so clear that orders *needed* chasing? I think not. Yet this is not untypical. There was efficiency at work here at one level, yet the total effect was wrong. Who was at fault? Was it the switchboard operator struggling to cope with the volume of calls? Or the manager supervising her? Or the sales or marketing department (which has final responsibility for customers)? The effect of producing similar efficiency, but coupled with the projection of a positive tone, would have been so much better.

Certainly such things should not go by default. No organization can afford to have inappropriate action taken simply because no one has thought about something, or thought adequately about it.

Attention to detail is important here. Communication is never as straightforward as we sometimes believe, and the obvious voice-only nature of telephone communication compounds the problem. If something hits the wrong note or is simply unclear, then we cannot see a puzzled expression at the other end of the line and may not so easily fine-tune an approach to correct matters and move on positively.

Like any communication, telephone contact must have a clear purpose. It is necessary to think before lifting the telephone to make a call – perhaps to plan and structure what will be done. Similarly, it is essential that taking incoming calls should not be viewed as being entirely responsive. You cannot just lift the phone and see what happens. We all need a clear idea of purpose.

- What is the call for?
- What are the caller's expectations?
- What are we trying to achieve?
- What impression should we give?
- How, exactly, must the call be handled: promptly (to project efficiency, corporate or individual personality)?
- What, as the receiver is replaced, should have been achieved?

So, our objective here is to review the process. To look at what can make this special kind of communication work for us, at the techniques involved and the skill with which the telephone must be handled.

8.3   An opportunity

Observation and experience suggest that, despite its importance, prevailing standards of telephone communication are not universally high. There are plenty of communications failures on the telephone; and sometimes they constitute total derailment. Whenever this is the case, there is an opportunity for those who get things right, and those who excel. And here it is one that demands no great cost or complication, just attention to detail, and the appropriate deployment of the key techniques involved.

The result can be powerful: callers – including customers – finding that their expectations are not just met, but exceeded. Impressions can be given that act not just to make the contact more pleasant and efficient, but also to enhance image and build business.

So, next we review some essential principles of this special form of communication.

8.4   The switchboard

Millions of times every day, telephones ring in offices all over the world. What happens next may be of little significance or of more, but it always matters. It may be a one-off contact, or part of a series of contacts that gradually form a view of the organization or person at the other end, but every time that ringing occurs it demands attention. Someone has to answer it, and that someone needs to deploy certain skills if the contact is to work and be successful, as does the caller.

The first point of contact is often a switchboard operator. Acting as an invisible receptionist, they set the scene for the ensuing contact and must do so effectively. To be effective they must be organized.

They need to know which calls go where, who does what, who backs up whom and where priorities lie. Is it more important to obtain an outside number for the managing director or attend to an incoming customer enquiry? They need to know something about the outside world, too. Who calls regularly? Whom do they speak to? How well do they know us? Do they like to be recognized? Operating the switchboard is no repetitive, parrot-like, function. It is important to sound lively and interested and to act promptly and efficiently.

There is a need for:

- answering *promptly*;
- providing a friendly greeting, such as 'Good morning'; having this come first avoids anything crucial being missed, because it can take the listener a moment to 'tune in' to the first words spoken and something may be missed;
- clearly *identifying* the organization reached;
- *offering assistance*: 'How may I help you?' (*may* is grammatically better than *can*, and such a phrase must be said carefully to avoid an insincere have-a-nice-day feel).

Assistance must then be offered in a friendly and understandable manner. It is useful to think about the kind of question that will most easily identify where a call should go – 'Is this a first enquiry?' – and once action is taken it must be efficient and accurate. People resent holding on for an age, unable to get back to the switchboard if an extension rings unanswered (though perhaps it should not do so), and alternatives must be offered helpfully – 'Can I get them to call you?'; 'Would you like me to find someone else who can help?' – if necessary. If there are delays, tell people exactly what is happening and always remember that waiting on the telephone seems very much longer than it is.

A good switchboard is a real asset to any organization, acting moment by moment to build image and smooth communications.

Panel 8.1 A warning

Ahead of the switchboard is increasingly arrayed an automated computer system to channel calls. What of these? Many would say simply, 'Avoid them,' believing that the human touch is always better. If they are necessary, however, it must be ensured that they are logical and work efficiently. They should offer a reasonable number of options (not too many so that going through them becomes interminable), and they should be in a sensible order, not least allowing someone who wants to get through and start a real conversation to do so, bypassing the automated system quickly.

Similarly, queuing systems should involve only a reasonable wait and plenty of information – 'You are second in the queue' – as you go along. Systems that offer to call you back (perfectly possible technically) are appreciated when the wait may be long. This is an area where many feel all that is necessary to improve things is the application of some common sense. Certainly, overcomplicated and obviously ill-thought-out systems can cause disproportionate annoyance.

Making a call may need some planning, but next we look at the need to respond appropriately when your telephone rings.

8.5 Taking a call

When the phone rings you may not know who it is calling, so the response must always assume that how you answer matters. If most of your calls are internal, you may be able to be a little more relaxed; if most are from customers, you need to act accordingly.

The basic rules include the need to carry out the following.

- Answer *promptly*.
- *Identify* yourself. It may be sufficient to say your name, and a fully stated name sounds best: 'Patrick Forsyth'. Or you may prefer something more: 'Good morning, this is Patrick Forsyth'. Or you may need something that includes a departmental or functional description, such as, 'Good morning, sales office, Mary Bolton speaking.' This may, in part, be a matter of policy and consistency around a department and should link neatly to whatever a switchboard says if calls are coming via it.
- *Hold the phone properly*. This may seem obvious, but it really does impede hearing if the handset is tucked under the chin or pushed aside as you reach for something. You must be clearly audible; hence the headsets increasingly issued to many people working largely on the telephone and used universally in call centres. Incidentally, such operations are important to those companies that use them, though problems can occur with the relationship between operating cost and efficiency (demonstrated by the – not always so efficient – recent relocations of such sections to developing countries such as India).
- *Decide whether to take the call or not*. Some calls, say from customers, may always need to be taken at once. But the telephone is obtrusive and, if you are busy and a colleague simply wants a word it may be acceptable to delay it: 'I'm just finishing an important draft; may I call you back in half an hour?' Alternatively, the call may need transferring – something that must always be explained and handled promptly.
- Adopt an *appropriate manner*. This is not a question of insincerity or acting, just that we want to emphasize certain traits more with some people than others – a nice friendly, but not overfamiliar, tone with customers, say, and so on.
- Once into the call, speak *clearly (and a fraction slower than normal)*. Certainly, do not gabble, and try to keep your thoughts organized. It is difficult talking instantly (because the phone has rung) and, as you cannot see the other person, thoughts can easily become jumbled as a result.

- It may help to *signpost* – to say something that tells the caller what is coming: 'Right, you want an update about the new brochure-design project. Let me go through it: the key things are probably the costs, the copy and the design – and the timing. Now, first, costs …' This helps both parties, giving the caller the opportunity to amend the list and giving you a list to keep in mind (note down?) and work through.

- Always *listen carefully*. It may be a voice-only medium, but it is two-way. Do not do all the talking, and make it clear you are listening by acknowledging points as they go along and *make notes as necessary*.

- Be *polite*. Of course, there could just be times you *need* to shout at someone (colleagues if not customers!), but more usually it is important to maintain reasonable courtesies and, with voice only, it can be easy to sound, say, abrupt when you are simply trying to be prompt.

- Be aware of *pauses*. If you say, 'Hold on a sec, I'll get the file,' remember that the pause seems long to the person waiting. Sometimes, it is better to suggest getting everything in front of you and phoning back. Or you can split what would otherwise seem a long pause into two, shorter, pauses by saying something like, 'Right, I've got the file, I'll just turn up the figures you want.' Note that this is especially important when someone is working with a computer in front of them and time taken to do something on screen needs explaining to a caller to explain an otherwise mysterious gap in the conversation (this also links to the next point).

- Have *the right information to hand*. Many calls are repetitive in nature. You can handle them more efficiently if you anticipate what information is needed and have it to hand (and in a form that is convenient to deal with on the phone, e.g. papers in ring binders lie flat and do not need a hand to hold them open when you are already trying to hold the handset and write notes).

- Be *careful of names*. People are sensitive about their names. Get them right – do so early on. Ask for the spelling or how to pronounce the name if necessary, and use them occasionally during the conversation. It is annoying if, as an outside caller, you are asked your name by the switchboard, a secretary and the person who handles the call and then, five minutes later, are asked 'What was your name again?'

- Technically, phone systems are now pretty good but if a *bad line* should really hinder communication it may be better to call again. Nowadays there is also the problem of mobile phones fading away or cutting out as conditions and perhaps location change. Thus, it is good practice always to say you are on a mobile at the start of a conversation when this is a possibility.

- Finally, *hang up last*. Now you cannot both do this! But, with something like a customer call, it is fine to be the first to initiate the end of the call or to say 'Goodbye' – not 'Bye-ee', incidentally – but, if you put the phone down last, it avoids the caller thinking of something else they want to say, finding only a dead line and feeling they have been left too hastily.

Such a list could be extended, and extended, no doubt; but the above captures the essentials and leaves space in this section for one other important basic point.

8.6 Projecting the 'personal/corporate personality'

There is more to good telephone manner than just being polite and saying 'Good morning' as if you mean it. The general tone you adopt sets the scene and gives people a clear impression of helpfulness, efficiency or whatever. But there are more powerful influences involved here.

People do not just take such communication at face value: they read between the lines. They ask themselves questions and believe that what they hear tacitly answers them. Who has not been surprised on meeting someone known only over the telephone, to find they are not at all as you imagined? Yet such initial views can seem so clearly right.

People ask, 'What kind of person is this? Would I like to know them? What will they be like to deal with in future?' Is what is happening so good that we make a note of their name and resolve to ask for them specifically next time? Or is the projected image offputting in some way: it is all too much trouble, it all takes a ridiculously long time, and they do not seem to know or be able to decide anything?

Both with internal and external contacts, those who know you and those who do not, the way you come over is vital; it matters to you and it matters to your organization. Consider the internal, then external, ramifications in turn.

8.6.1 Internal image

This is not the place for a treatise on your personal style; suffice to say that what you do affects how you are seen. It is the same in other areas. For example, if someone looks a mess, their desk looks as if a bomb had hit it and they always miss every deadline they are set, then they may not seem the most reliable choice for an interesting project or for promotion.

Just because you are not visible when you speak on the telephone, it does not mean you are not sending out signals. You are. And you have some choice as to what they are. Do you want to come over as bright and efficient, or surly and unhelpful? How you act, what you say and the way you use your voice – all influence this and are worth considering. Deploy the appropriate signs and you will feel the benefit over time.

8.6.2 External image

Here the implications are more complex because people judge the organization *through* an individual. They may literally see them as personifying it. Action here thus starts with clarity of purpose. How does your organization want to be seen? This needs spelling out, not simply in terms of 'good things' – efficient, high quality, etc. – but in terms of values and feelings. Are you a caring, an innovative or an exceptional organization? And, if so, in what way?

Then you need to think about how this statement should be reflected in the manner that comes over on the telephone. For example, a caring company (in healthcare, say, or your own organization?) must surely have time for people, so unseemly haste on the telephone might sit awkwardly with that image. Similarly, if your contacts are all at a high level, then some visible respect might be appropriate; while other business relationships come to thrive on a chatty informality (while always remaining efficient).

8.7   A telephone 'handshake'

Overall, most of us do not like to deal at arm's length. We like to get to know people, especially if we are to deal with them regularly. On first meeting we are sometimes conscious of a drawing together as we begin to get the measure of people – even shaking hands is part of such a relationship developing.

On the telephone it seems that things are, by definition, at arm's length. Perhaps it defines good telephone contact well when we are able to say that the style of the contact avoids this feeling. People who make others feel they know something about them, who come over as likable – and certainly easy and pleasant to deal with – and who can, over time, create and maintain real relationships are always a pleasure to speak with, whoever instigates the contact.

Whatever rules, principles or good intentions there may be, the whole process of telephone communication is dependent on one thing: the voice and how it is used.

Panel 8.1   Messages

What would you want if you had missed a call while reading this chapter? A clear message, without a doubt. Messages are important: just one lost message can cause chaos. The world is full of people puzzling over tiny bits of paper with indecipherable messages on them: 'Brown (Browne?) rang – told him you would call.'

Which Brown or Browne? What did he want? How urgent was it? What was it about? When did the message come in? And who took it? A good, clear message form is essential – preferably important-looking (some organizations have coloured message forms, often full A4 size, to stand out), and certainly designed to take and relay to you *exactly* the information *you* want (for instance, making a useful distinction between 'action taken' and 'action promised'). Make sure whatever you use is tailored to work as you want.

8.8   Making the voice work for you

None of the techniques referred to so far will work well without consideration of that most important element of telephone manner: the voice. Not only is the voice an important element, but it has to act alone. By definition, telephone communication is voice-only. This makes for some difficulty (remember the job of describing a necktie mentioned in the Preface and again in Chapter 2).

The telephone is itself apt to make communication more difficult than it would be face to face. When we speak on the telephone, the telephone distorts the voice somewhat (more so if there is a bad line), it seems to exaggerate the rate of speech and it heightens the tone.

Basics first: you must talk directly into the mouthpiece in a clear, normal voice (some women may need to pitch the voice just somewhat lower than normal to avoid any squeakiness). It is surprising how many things can interfere with the simple process of speaking into the mouthpiece, for example:

- trying to hold a file open;
- trying to write;
- allowing others in the office to interrupt;
- eating or smoking (the latter now banned from offices).

All the above have an adverse effect; you may be able to add to the list. Some just need avoiding: do not eat or smoke when on the telephone. Others need organization: for example, information in a ring binder will lie flat and avoid the necessity of trying to hold a file open, write *and* hold the handset. All it takes is a little thought and organization to prevent such things diluting the effectiveness of communication just a little.

8.9   A wrong impression

One more point is worth noting before we get into the positive aspects, and that is the danger of humour or irony when you cannot visually judge the reaction. It is not suggested that you become exclusively po-faced, but you do need to be careful. Recalling the face of an aggrieved customer in a training film about the use of the telephone suggests an illustration. Someone (less organized than they should be) is desperately trying to recall a name given over the telephone earlier in the conversation: 'Rolls, isn't it?' they say and are corrected: 'Bentley.' An attempt at humour over the confusion of two names, both used for cars, then clearly fails and makes matters worse.

Be doubly careful about anything that might be inappropriate for someone of different background or ethnic origin, or who does not speak good English.

8.10   Voice and manner

Here too the details matter (and it might be worth reminding yourself how you sound by recording your voice – it's not as you normally hear it). The following, in no particular order of priority, are all important:

8.10.1   *Speak slightly more slowly than normal*

You do not need to overdo this and slow down so that you appear to be half a-s-l-e-e-p. But pace is important. A considered pace will be more likely to allow things to be made clear, and misunderstandings avoided. It allows the listener to keep up, particularly, for example, when it is clear they may be wanting to make a note; slow down especially for that.

In addition, too rapid-fire a delivery can sound glib and promote a lack of trust. It is important not to sound like a dodgy second-hand-car dealer, who will always go at a rate that precludes easy interruption. Also, rapidity can allow key points to be missed. As an example, who has not had a message left on an answering machine saying 'Please ring 01621 burble, burble'? There is a lesson here.

8.10.2   Make inflection work for you

Inflection is what makes, for example, a question mark stand out at the end of a sentence, and also what gives variety and interest to the way you speak. It is important that intended inflection be noticed.

8.10.3   Smile

You do not need to hold a fixed grin, but, even though a smile cannot be seen, a pleasant smile produces a pleasant tone and this does make for the right sound. A warm tone of voice produces a feeling that the speaker is pleasant, efficient, helpful and, most important, interested in the person at the other end.

There are many situations (with customers, for one) when enthusiasm is important. This has to be heard; and it is about the only good thing in life that is contagious.

8.10.4   Get the emphasis right

It is necessary to get the emphasis right in terms of both words – '*This* is really important'; or, 'This is *really important*' – and in terms of the part of the message to which the listener must pay most attention.

For example, often we may recall struggling to note a barrage of detail coming at us over the phone, when the person concerned suddenly says something like, 'The details don't really matter; when you come through to us next time just quote the following reference …This will get you through at once.' Clearly, it would have been better to say that first – another useful principle.

8.10.5   Ensure clarity

It is no good sounding pleasant if what you say cannot be understood. Be clear and be particularly careful about: names, numbers (you do not want to allow 15 per cent discount to be confused with 50 per cent, for instance), and sounds that can be difficult to distinguish: Fs and Ss for instance. Just good, thoughtful articulation helps here.

An important detail is worth emphasizing. Find a way of doing it that works. For example, for a postcode ending 7BB, it may be best over the telephone to add something like 'That's B for butter', in case 'P' is 'heard' instead.

Exercise some care if you have an accent (say a regional accent). You have no reason to apologize for it, but may need to bear in mind that some elements of it will not be as clear to others as they are to you. That said, though, some organizations favour the character a regional accent lends to a telephone transaction, which might otherwise seem impersonal. The telephone banking organization First Direct is one such.

8.10.6   Be positive

This is especially important where an impression of efficiency is key. Avoid saying 'possibly', 'maybe' and 'I think', when the expectation is that you should give definitive information: 'It is …' (But do not waffle. If you do not know, say so – you can always offer to get back to people.)

8.10.7   Be concise

Most of the people you will speak with, in a business context, expect and appreciate it if you value their time. This especially means that convoluted descriptions need to be thought about in advance, and made concise, yet precise.

Be careful with the social chat. It is often liked by regular contacts, but there can be a thin line between its being a pleasure to hear you again, and its becoming a time waster.

8.10.8   Avoid jargon

Jargon is professional shorthand, and can be very useful – in its place. But you need to be sure of what another person understands and select the level of jargon to be used accordingly. Otherwise, you can find you are blinding people with science, as it were, and some – not wanting to appear foolish by asking – may allow meaning to be diluted.

For example, beware of *company* jargon (abbreviations of a department, process or person, perhaps) and of *industry* jargon (technical descriptions of products and processes), and even of *general phrases* that contain an agreed internal definition that is not immediately apparent to an outsider: such as 'good delivery'. What is 24-hour service, other than not sufficiently well defined? You can probably think of many more examples, some close to home. Remember, on the telephone, you cannot see signs of confusion.

8.10.9   Be descriptive

Good description can add powerfully to any message. There is all the difference in the world between saying that something is 'smooth as silk' and describing it as 'sort of shiny'. Things that are inherently difficult to describe can create powerful impact if a well-thought-out description surprises by its eloquence.

This is especially true of anything where the phraseology is not just clear but novel. For example, the sales executive of a hotel arranging a room for a training session described a chosen layout to me – a U-shape – as 'putting everyone in the front row'. Well said! That's just what it is.

Conversely, beware of bland descriptions that impart minimal meaning. This means no company's product is merely 'quite nice', and something that is 'user-friendly' nowadays fails to differentiate itself from anything else.

8.10.10   Use gestures

Yes, of course, they cannot be seen. But they may make a difference to how you sound, contributing to a suitable emphasis, for instance. Be careful, of course: you have to hang onto the phone and not knock everything off the desk! Some calls may be easier to make standing up. Really! Try it. This applies to such things as chasing a debtor or anything needing a fair bit of assertiveness to be displayed. It creates an antidote to the 'Sorry to worry you, but I wonder if perhaps ...' approach.

8.10.11   Adopt the right tone

In most circumstances you want to be friendly without being flippant; you want to sound courteous (*always* with customers) and usually you want to tailor your style to the circumstances, consciously deciding whether to produce a note of respect, a feeling of attention to detail or whatever. Getting this right is what produces a good telephone 'handshake' feeling.

8.10.12   Sound yourself

Be yourself. And certainly avoid adopting a separate, contrived 'telephone voice'; it does not tend to work and is difficult to sustain.

All of the above are things that can be consciously varied. Some – clarity, for instance – may need experiment, rehearsal and practice. But together they combine to produce a satisfactory manner. The effect is cumulative, and this works both ways. It means that any shortfalls begin to add up, eventually diluting the overall power of what is done. Equally, the better you work in all these areas, the more the effects combine to create a satisfactory overall impression and style.

How your voice sounds goes logically with the way you use language, so we turn next to a few points under this heading.

8.11   Use of language

Several of the points above touch on language as much as voice: descriptiveness for one. The point has also already been made that you should be yourself. So avoid 'office-speak'. A few examples are sufficient to make the point; do not say:

- 'at this moment in time' when you mean 'now';
- 'due to the fact that' ('because');
- 'I am inclined to the view that' ('I think').

And do not overdo the 'we' – who is this corporate 'we' for goodness' sake? Make things personal – 'I will ensure that' sounds totally committed. Refer to people by name: 'Mary Brown in accounts will ...' is much better than 'Someone will ...'

Although grammatical perfection is not essential in conversation, it is good to avoid those things that are wrong and irritate. For example, someone incorrectly adding several additional comparatives or superlatives to the word *unique*, and saying things like 'more unique' or 'the most unique' (and even 'very unique') will annoy, and therefore be inclined to detract from any good sense being talked.

Watch also for habits which can introduce an annoying or incorrect note: for example, ending every other sentence with 'right' or 'OK', or starting with a superfluous use of 'basically'. Somehow, in voice-only communication, these things stand out – and are to be avoided.

8.12   Listening

However well you speak and however well you get your point over, no telephone conversation can be a monologue. You need to generate feedback and the first step to this is, not unnaturally, to listen. There is an old saying that people have two ears and one mouth for a good reason. Certainly, we should always remember that listening is just as important as speaking.

Good communication demands good listening skills, and this is especially vital on the telephone, when there are few other signs. Not only does it give you more information, but others like it. But you need to work at it. The details will not be repeated here, but Paragraph 3.2 in Chapter 3 describes listening as an 'active process' and may be worth referring back to before we move on.

Certainly, good listening skills are a sound foundation for any sort of communication, and on the telephone something like a distraction cannot be seen and may not be appreciated at the other end.

8.13   Creating a dialogue

Two-way communication does not demand merely talking one at a time and listening in between. Creating a dialogue is something you need to work actively at. Here are some examples of how that might be achieved.

- **Talk with people, not at them**: It may help to form a mental picture of the person at the other end. Certainly, treating them like a disembodied voice does not create the right impact.
- **Maintain a two-way flow**: Do not interrupt. Make sure that, if they are talking at some length, they are sure you are still there and listening: 'Right …' And flag what you are going to do to make your intentions clear: 'Good, I have those details, now perhaps I can just set out …'
- **Do not jump to conclusions**: This should not be done for whatever reasons. It may be that you do know what is coming, but if you make unwarranted assumptions it can cause problems.
- **Give the feeling that things are being well handled**: The dialogue should not just flow: it should actively appear to sort or deal with things as necessary. The

whole manner and structure need to get to the point and clearly be doing whatever is necessary to sort something.

The people you speak to should feel you want to talk to them, that you will let them have their say, and that you listen. At the same time, there is a necessity in many conversations, from both parties' points of view, to keep calls short and businesslike.

8.14   Projecting the right image

Every time you pick up the telephone and speak, someone will form an impression of you and through you of the department/organization for which you work. This needs some thought. What do you *want* them to think of you? The answer will depend on the person and the circumstances.

If you are talking to customers, there is quite a list of characteristics one doubtless wants to get over: courtesy, promptness, expertise, efficiency, good advice and many more. But a similar situation in fact exists whomever it is you speak to, whether they are internal or external.

This was touched on earlier, and the point will not be revisited here, except to say that the only channel to express whatever aspects are important here is the voice.

So, with general points about how you come over in mind, we can perhaps end this chapter by mentioning that the content and purpose of telephone calls range widely. This chapter deals with the basic techniques and illustrates the surprising complexity involved; it may usefully be read in context with other topics in this book. For instance, you may need to be, say, persuasive – in such cases the appropriate techniques for that need deploying alongside those for the telephone.

Perhaps, the most important calls any organization gets are those from its customers (though there may be people you want or need to regard as 'internal customers'). But all calls have to be handled just right if people are to be kept happy and satisfied and their business is to be confirmed or continue.

8.15   Summary

Overall the key issues are to:

- respect the medium – voice only restricts what can be done;
- use the telephone in a considered (and sometimes prepared) way to avoid the mouth running ahead of the thinking;
- make sure that voice and use of language – the only things that can affect clarity and image – are used to good effect.

Chapter 9
The nature of the written word

In a busy business life, writing anything can be a chore. There are surely more important things to be done: people to meet, decisions to be made, action to be taken. Yet all of these things and more can be dependent on written communication. A letter or memo may set up a meeting; a report may present a case and prompt a decision; a proposal may act persuasively to make sure certain action is taken or a particular option is selected.

But reading business papers can be a chore also, and they will not achieve their purpose unless they are read and understood and do their job well enough to prompt the reader to action.

Business writing must *earn* a reading.

You are probably both a reader and a writer of business documents. Consider writing with your reader's hat on for a moment. Do you read everything that crosses your desk? Do you read every word of the things you do read? Do you read everything from the first word through in sequence, or do you dip into things? Almost certainly, the answers make it clear that not all writing is treated equally. Some documents are more likely to be read than others. Of course, some subjects demand your attention. Who ignores a personal note from the managing director? But the fact that some things have to be read does not make their reading any easier or more pleasurable.

Good writing, which means, not least, something that is easy to read and understand, will always be likely to get more attention than sloppy writing. Yet we all know that prevailing standards in this area are by no means universally good.

Why is this? Maybe it is education – or lack of it. Often school assists little with the kind of writing we find ourselves having to do once we are in an organization. Maybe it is lack of feedback – perhaps managers are too tolerant of what is put in front of them. If more of it were rejected and had to be rewritten, then more attention might be brought to bear on the task.

Habits are important here, too. We all develop a style of writing and may find it difficult to shift away from it. Worse, bad habits may be reinforced by practice. For example, the ubiquitous standard document can often be used year after year with no one prepared to say scrap it, even if they notice how inadequate it is.

9.1 A fragile process

We can all recognize the really bad report, without structure or style, but with an excess of jargon and convoluted sentences, which prompts the thought, 'What is it trying to say?' But such documents do not have to be a complete mess to fail in their purpose. They are inherently fragile. One wrongly chosen word may dilute understanding or act to remove what would otherwise be a positive impression made.

Even something as simple as a spelling mistake – and, no, computer spellcheckers are not infallible – may have a negative effect: I once saw work lost by a consultant who spelt the name of a company as *diary* rather than *dairy*. As a very first rule to drum into your subconscious, check, check and check again.

(I treasure the computer manual that states, 'The information presented in this publication has been carefully for reliability.' No one is infallible, but I digress.)

Whether the cause of a document being less good than it should be is major or minor, the damage is the same, and shows that the quality of writing matters.

9.2   A major opportunity

Whatever the reasons for poor writing may be, suffice to say that, if prevailing standards are low, then there is a major opportunity here for those who better that standard. More so for those who excel. Bad documents might just come back to haunt you later.

So, business writing is a vital skill. There may be a great deal hanging on a document's ability to do the job it is intended to do – a decision, a sale, a financial result, a personal reputation. For those who can acquire sound skills in this area, very real opportunities exist. The more you have to write, and the more important the documents you create, then the truer this is. Quite simply, if you write well, then you are more likely to achieve your business goals.

This point cannot be overemphasized. One sheet of paper may not change the world, but – well written – it can influence many events in a way that affects results and those doing the writing.

And you *can* write well. We may not all aspire to or succeed in writing the great novel, but most people can learn to turn out good business writing. Writing that is well tailored to its purpose and likely to create the effect it intends. This chapter and the next two review some of the approaches that can make business writing easier, quicker – a worthwhile end in itself – and, most important, more likely to achieve its purpose.

Good business writing need not be difficult. It is a skill that can be developed with study and practice. Some effort may be involved, and certainly practice helps, but it could be worse. Somerset Maugham is quoted as saying, 'There are three rules for writing the novel. Unfortunately, no one knows what they are.' Business writing is not so dependent on creativity, though this is involved, and it *is* subject to certain rules. Rules, of course, are made to be broken. But they do act as useful guidelines and can therefore be a help. Here we review how to go about the writing task and, in part, when to follow the rules and when to break them.

9.3   What makes good business writing?

Despite predictions about the 'paperless office', offices seem as surrounded by – submerged in? – paper as ever. Indeed, as documentation is essentially only a form of communication, this is likely to remain so. However a case is presented, even if there is no paper – as with something sent via email, for example – it has to be written.

With no communication, any organization is stifled. Without communication, nothing much would happen. Communication – good communication – should oil the wheels of organizational activity and facilitate action. This is true of even the simplest memo, and is certainly so of something longer and more complex, such as a report.

Communication is – inherently – inclined to be less than straightforward. If this is true of tiny communications, how much more potential for misunderstanding does a 25-page report present? And with written communication the danger is that any confusion lasts. There is not necessarily an immediate opportunity to check (the writer might be a hundred miles away), and a misunderstanding on Page 3 may skew the whole message taken from an entire report.

9.3.1   Serious, and very serious

Once something is in writing, any error that causes misunderstanding is made permanent, or at least set in place for a while. The dangers of ill-thought-out writing vary.

- **It may be wrong, but still manage to convey its meaning**: In the bedrooms of one hotel, for instance, there is a notice saying, 'In the interest of security, please ensure that your bedroom door is fully closed when entering or leaving your room.' It may amuse – and be a good trick if you can do it – but it will probably be understood. No great harm done, perhaps, though in a service business any fault tends to highlight the possibility of other, more serious, faults.
- **It may try too hard to please, ending up giving the wrong impression**: In one Renaissance Hotel there are signs on the coffee shop tables that say, 'COURTESY OF CHOICE: The concept and symbol of "Courtesy of Choice" reflect the centuries-old philosophy that acknowledges differences while allowing them to exist together in harmony. "Courtesy of Choice" accommodates the preferences of individuals by offering both smoking and non-smoking areas in the spirit of conviviality and mutual respect.' An absurd overpoliteness just ends up making the message sound rude – this restaurant has both smoking and non-smoking areas and if you find yourself next to a smoker, tough. It does matter.
- **It may be incomprehensible**: A press release is an important piece of writing. One, quoted in the national press recently, was sent out by the consulting group Accenture. The item commented that Accenture envisioned, 'A world where economic activity is ubiquitous, unbounded by the traditional definitions of commerce and universal.' Er, yes – or, rather, no. The newspaper referred not to the content of the release, only to the fact that it contained a statement so wholly gobbledegook as to have no meaning at all. It is sad when the writing is so bad that it achieves less than nothing.

You could doubtless extend such a list of examples extensively. The point here is clear: it is all too easy for the written word to fail. All the above were probably the subject of some thought and checking; but not enough. Put pen to paper and you step onto dangerous ground.

So, the first requirement of good business writing is clarity. A good report needs thinking about if it is to be clear, and it should never be taken for granted that understanding will be automatically generated by what we write.

It is more likely that we will give due consideration to clarity, and give the attention it needs to achieving it, if we are clear about the purpose of any report we may write.

9.4   Why are we writing?

Exactly why anything is written is important. This may seem self-evident, yet many reports, for instance, are no more than something 'about' their topic. Their purpose is not clear. Without clear intentions the tendency is for the report to ramble, to go round and round and not come to any clear conclusion.

Documents may be written for many reasons. For example, they may intend to inform, motivate or persuade – any of the intentions listed earlier – and often more than one intention is aimed at, and different messages or emphasis for different people add further complexity.

9.4.1   Reader expectations

If a document is to be well received, then it must meet certain expectations of its readers. Before going into these, let us consider generally what conditions such expectations. Psychologists talk about what they call *cognitive cost*. This is best explained by example. Imagine you want to programme the video recorder. You want to do something that is other than routine, so you get out the instruction book. Big mistake. You open it (try this, you can open it at random) and the two-page spread shouts at you, 'This is going to be difficult!' Such a document has what is called a *high cognitive cost*, rather than appearing inviting; even a cursory look is offputting.

People are wary of this effect. They look at any document almost *expecting* reading it to be hard work. If they discover it looks easier and more inviting than they thought (a low cognitive cost), then they are likely to read it with more enthusiasm. What gives people the feeling, both at first glance and as they get further into it, that a document is not to be avoided on principle? This is spelt out in the next paragraph.

9.4.2   Reader preference

In no particular order, the following are some of the key factors readers like.

- **Brevity**: Obviously, something shorter is likely to appear to be easier to read than something long, but what really matters is that a document should be of an appropriate length for its topic and purpose. Perhaps the best word to apply is *succinct* – to the point, long enough to say what is necessary and no more. A report may be ten pages long or fifty, and still qualify for this description.

- **Succinctness**: This makes clear that length is inextricably linked to message. If there is a rule, then it is to make something long enough to carry the message – then stop.
- **Relevance**: This goes with the first two. Not too long, covering what is required, and without irrelevant content or digression. (Note: comprehensiveness is *never* an objective. If a report, say, touched on absolutely everything, it would certainly be too long. In fact, you always have to be selective. If you do not say everything, then everything you do say is a choice – you need to make good content choices.)
- **Clarity**: Readers must be able to understand it. And this applies in numbers of ways. For example, it should be clearly written (in the sense of not being convoluted), and use appropriate language – you should not feel that, as an intended reader, you have to look up every second word in a dictionary.
- **Precision**: Saying exactly what is necessary and not constantly digressing without purpose.
- **'Our language'**: In other words, using a level and style of language that is likely to make sense to the average reader, and that displays evidence of being designed to do so.
- **Simplicity**: Avoiding unnecessary complexity.
- **Helpful structure**: So that it proceeds logically through a sequence that is clear and makes sense as a sensible way of dealing with the message
- **Descriptiveness**: Of which more anon. Here, suffice it to say that if there is a need to paint a picture it must be done in a way that gets that picture over.

All these have in common that they can act to make reading easier. Further, they act cumulatively: that is, the more things are right in each of these ways, the clearer the meaning will be. If the impression is given that attention has *actively* been given to making the reader's task easier, so much the better.

Such factors are worth personalizing to the kind of people to whom you must write. Whether this is internal (colleagues, perhaps), or external (customers, say, or collaborators), you need to be clear what your communications have to do and what kinds of expectations exist at the other end. For example, a technical person may have different expectations from a layman, and may be looking to check a level of detail that must exist and be clearly expressed for a report to be acceptable to them.

9.4.3   *The readers' perspective*

It follows logically from what has been said in this chapter so far that good business writing must reflect the needs of the reader. Such writing cannot be undertaken in a vacuum. It is not simply an opportunity for the writer to say things as they want. Ultimately, only readers can judge a document to be good. Thus their perspective is the starting point and as the writer you need to think about who the intended readers are, how they think, how they view the topic of the report, what their experience to date is of the issues, and how they are likely to react to what you have to say. This factor links to preparation, which is dealt with in depth separately.

9.5   Powerful habits

Habit, and the ongoing pressure of business, can combine to push people into writing on 'automatic pilot'. Sometimes, if you critique something that you wrote, or that went out from your department, you can clearly see something that is wrong. A sentence does not make sense, a point fails to get across or a description confuses rather than clarifies. Usually the reason this has occurred is not that the writer really thought this was the best sentence or phrase and got it wrong. Rather, it was because there was inadequate thought of any sort – or none at all.

Habits can be difficult to break and the end result can be a plethora of material moving around organizations couched in a kind of gobbledegook, or what some call 'office-speak'.

9.6   Earning a reading

The moral here is clear. Good writing does not just happen. It needs some thought and some effort (and some study, with which this book aims to assist). The process needs to be actively worked at if the result is going to do the job you have in mind, and do it with some certainty.

But good habits are as powerful as bad. A shift from one to another is possible and the rewards in this case make the game very much worth the candle. Think what good writing skills can achieve.

9.6.1   The rewards of excellence

Consider the example of reports. They can influence action. But they also act to create an image of the writer. Within an organization of any size, people interact through communication. They send each other memos, they sit in meetings and on committees, they chat as they pass on the stairs, or share a sandwich at lunchtime; and all of this sends out signals. It tells the world, or at least the organization, something about them. Are they knowledgeable, competent, expert, easy to deal with, decisive? Would you take their advice, follow their lead or support their cause?

All the different ways in which people interrelate act together, cumulatively and progressively, to build up and maintain an image of each individual. Some ways may play a disproportionate part, and report-writing is one such. There are two reasons why this effect is important.

- Reports, unlike more transient means of communication, can last. They are passed around, considered and remain on the record – more so if they are about important issues.
- Because not everyone can write a good report, people can be impressed by a clear ability to marshal an argument and put it over in writing.

Thus, reports represent an opportunity, or in fact two opportunities. Reports – at least, good ones – can be instrumental in prompting action, action you want, perhaps.

They are also important to your profile. They say something about the kind of person you are and how you are to work with. In a sense there are situations where you want to make sure certain personal qualities shine through. A case may be supported by its being clear that it is presented by someone who gives attention to details, for instance.

In the longer term, the view taken of someone by their superiors may be influenced by their regularly reading what they regard as good reports. So, next time you are burning the midnight oil to get some seemingly tedious report finalized, think of it as the business equivalent of an open goal and remember, it could literally be affecting your chances of promotion!

9.7   A significant opportunity

Reports demand detailed work. Their preparation may, on occasion, seem tedious. They certainly need adequate time set aside for them. But, as the old saying has it, if a job is worth doing, it is worth doing well. It may take no more time to prepare a good report than it does to prepare a lacklustre one. So too for any document. Indeed, the next chapter contends that a systematic approach can speed up your writing.

If whatever you write is clear, focused and set out so as to earn a reading, then it is more likely to achieve its purpose; it is also more likely to act positively to enhance the profile of the writer. Both these results are surely worthwhile. But the job still has to be done, the words still have to be got down on paper, and, if you are faced with a blank sheet (or, these days, screen), this can be a daunting task (writing a book of this length certainly qualifies me to say that!). Go about it in the right way and it does become possible.

9.8   Summary

- Remember, communication has inherent dangers; clear communication needs to be well considered.
- Any document will achieve its purpose only if the writer is clear in their mind what they are seeking to achieve.
- The reader is more important than the writer; write for others not for yourself
- Beware old habits and work to establish good ones.
- Written documents are potentially powerful tools – powerful in action terms, and powerful in contributing to personal profile.

Chapter 10
The writing process: what to say and how to say it

If you undertake to engender a totality of meaning which corresponds with the cognition of others seeking to intake a communication from the content you display in a report, there is a greater likelihood of subsequent action being that which you desire.

You're right: that is not a good start. If I want to say, 'If you write well, people will understand and be more likely to react as you wish,' then I should say just that. But it makes a good point with which to start this chapter. Language and how you use it matter. Exactly how you put things has a direct bearing on how they are received; and that in turn has a direct bearing on how well a report succeeds in its objectives.

10.1 The difference language makes

It is clear language that makes a difference. But this is a serious understatement, for language can make a very considerable difference. And it can make a difference in many different ways, as this chapter will show.

How you *need* to write must stem as much as anything from the view your intended readers have of what they want to read. Or in some cases are prepared to read, because – be honest – reading some business documents can be something of a chore; maybe even some of those you write.

Consider four broad elements first. Readers want documents to be *understandable*, *readable*, *straightforward* and *natural*. Let us look at each in turn.

10.1.1 Understandable

Clarity has been mentioned already. Its necessity may seem to go without saying, though some, at least, of what one sees of prevailing standards suggests the opposite. It is all too easy to find everyday examples of wording that is less than clear. A favourite is a sign you see in some shops: 'Ears pierced, while you wait'. There is some other way? Maybe there has been a new technological development.

Clarity is assisted by many of the elements mentioned in this chapter, but three factors help immensely.

1. **Using the right words**: For example, are you writing about *recommendations* or *options*, about *objectives* (desired results) or *strategies* (routes to achieving objectives), and when do you use *aims* or *goals*?

2. **Using the right phrases**: What is '24-hour service' exactly, other than not sufficiently specific? Ditto 'personal service'. Is this just saying that people do it? If so, it is hardly a glimpse of anything but the obvious. Perhaps it needs expanding to explain the nature, and perhaps excellence, of the particular service approach.

3. **Selecting and arranging words to ensure your meaning is clear**: For example, saying, 'At this stage, the arrangement is ...' implies that later it will be something else when this might not be intended. Saying 'After working late into the night, the report will be with you this afternoon', seems to imply (because of the sequence and arrangement of words) that it is the report that was working late.

Even changing a word or two can make a difference. Saying something is 'quite nice' is so bland that, if applied to something that is 'hugely enjoyable' it understates it so much as to be almost insulting. The emphasis may be inadequate but at least the word *nice* makes it clear that something positive is being said. Blandness is certainly to be avoided. It is unlikely to add power to your presentation, but choosing the wrong word is another matter. That might confuse, upset – or worse. The following examples are designed to show the danger.

- *Continuous* (unbroken or uninterrupted); *continual* (repeated or recurring). A project might be continuous (in process all the time), but work on it is more likely to be continual (unless you never sleep).
- Are you *uninterested* in a proposal or *disinterested* in it? The first implies you are apathetic and care not either way; the latter means you have nothing to gain from it.
- Similarly *dissatisfied* and *unsatisfied* should not be confused. Disappointed or needing more of something, respectively.
- You might want to do something *expeditious* (quick and efficient), but saying it is *expedient* might not be so well regarded, as it means only that something is convenient (not always a good reason to do anything).
- *Fortuitous* implies something happening accidentally; it does not mean fortunate.
- If you are a *practical* person, you are effective; if something is *practicable* it is merely possible to do; and *pragmatic* is something meant to be effective (rather than proven to be).

One wrong word may do damage. More, particularly when closely associated, quickly create nonsense: 'This practicable approach will ensure the practicable project will be continuous; it is fortuitous that I am uninterested in it and I am sure I will not be unsatisfied to see it start.'

Of course, no inaccurate use of language will help you put a message over well even if it only annoys rather than confuses: for example, saying 'very unique' when 'unique' means unlike anything else and cannot be qualified in this way; writing '12 noon' when 'noon' tells you everything you need to know; or talking about an 'ATM machine' when the *M* stands for 'machine' (a machine machine?). Then we see the ubiquitous 'PIN number' and 'HIV virus'. Some care, maybe even some checking or study, may be useful.

10.1.2   *Readable*

Readability is difficult to define, but we all know it when we experience it. Your writing must flow. One point must lead to another; the writing must strike the right tone, inject a little variety; and, above all, there must be a logical, and visible, structure to carry the message along. As well as the shape discussed in the previous chapter, the technique of 'signposting' – briefly flagging what is to come – helps in a practical sense to get the reader understanding where something is going. It makes them read on, content that the direction is sensible (this section starts just that way, listing points to come, of which 'readable' forms the second subsection). It is difficult to overuse signposting and it can be utilized at several levels within the text.

10.1.3   *Straightforward*

In a word (or two) this means *simply put*. Follow the well-known acronym KISS – 'Keep It Simple, Stupid'. Thus use the following.

- **Short words**: Why 'elucidate' something when you can 'explain'? Why 'reimbursements' rather than 'expenses'? Similarly, although 'experiment' and 'test' do have slightly different meanings, in a general sense 'test' may be better; or you could use 'try'.
- **Short phrases**: Do not say 'at this moment in time' when you mean 'now', or 'respectfully acknowledge' something, a suggestion perhaps, when you can simply say 'thank you for …'
- **Short sentences**: Having too many overlong sentences is a frequent characteristic of business reports. Short ones are good. However, they should be mixed in with longer ones, or reading becomes rather like the action of a machine gun. Many reports contain sentences that are overlong, often because they mix two rather different points. Break these into two and the overall readability improves
- **Short paragraphs**: If there are plenty of headings and bullet points it may be difficult to get this wrong, but keep an eye on it. Frequent and appropriate breaks as the message builds up do make for easy reading.

10.1.4   *Natural*

In the way that some people are said, disparagingly, to have a 'telephone voice', so some write in an unnatural fashion. Such a style may just be old-fashioned or bureaucratic. However, it could be made worse by attempts to create self-importance, or to make a topic seem weightier than it is. Just a few words can change the tone: saying 'the writer' may easily sound pompous, for instance, especially if there is no reason not to say 'I' (or 'me').

The moral here is clear and provides a guideline for good writing. Business documents do need some formality, but they are, after all, an alternative to talking to people. They should be as close to speech as is reasonably possible. It is not suggested that you overdo this, either by becoming too chatty or by writing, say, 'won't' (which you might acceptably say), when 'will not' is genuinely more suitable. However, if you compose what you write much as you would say it and then tighten it up, the end result is often better than when you set out to create something that is 'formal business writing'.

The four factors above have wide influence on writing style, but they do not act alone. Other points are important, and link to reader expectations.

Writing needs to be based very much on what people say they want in what they read. Just to headline this (without repeating detail from the last chapter) this means: brief, succinct, relevant, precise, clear, and in 'our' language.

10.2   Readers' dislikes

Readers also have hopes that what they must read will *not* be any of these.

- **Introspective**: It is appropriate in most business documents to use the word 'you' more than 'I' (or 'we', 'the company', 'the department' etc.). Thus saying, 'I will circulate more detailed information soon' may be better phrased as, 'You will receive more information [from me] soon.' More so, perhaps, if you add a phrase such as, '... so that you can judge for yourselves'. This approach is especially important if there is persuasion involved.

- **Talking down**: 'As an expert, I can tell you this must be avoided; you must never ...' Bad start – it sounds condescending. You are likely to carry people with you only if you avoid this kind of thing. As a schools broadcast on radio put it, 'Never talk down to people, never be condescending. You *do know* what "condescending" *means*, don't you?' Enough said.

- **Biased**: At least where it intends not to be. A manager writing something to staff setting out why he/she thinks something is a good idea, and then asking for the staff's views, may prompt more agreement than is actually felt. If views are wanted, then it is better simply to set something out and ask for comment, without expressing a firm personal view in advance.

- **Politically incorrect**: There is considerable sensitivity about this these days that should be neither ignored nor underestimated. As there is still no word that means 'he or she', some contrivance may be necessary in this respect occasionally. Similarly, choice of words needs care. One might be pulled up these days for using the expression 'manning the office'. If you meant who was on duty at what times, rather than anything to do with recruitment or selection (which the usually suggested alternative of 'staffing' seems to imply), this might seem somewhat silly. But, if it matters, it matters, and, while the way you write should not become awkward or contrived to accommodate such matters, some care is certainly necessary.

There is a considerable amount to bear in mind here. The focus must be on the reader throughout. However, you must not forget your own position as the writer: there are things here also that must be incorporated into the way you write.

10.3   The writer's approach

Every organization has an image. The only question is whether this just happens, for good or ill, or if it is seen as something actively to create, maintain and make positive. Similarly, every report or proposal you write says something about you. Whether you like it or not, this is true. And it matters. The profile wittingly or unwittingly presented may influence whether people believe, trust or like you. It may influence how they feel about your expertise, or whether they can see themselves agreeing with you or 'doing business' with you.

Your personal profile is not only an influence in your job, one that links to the objectives you have, but it also potentially affects your career. Surely it is unavoidable that, given the profusion of paperwork in most organizations, what you write progressively typecasts you in the eyes of others – including your boss – as the sort of person who is going places, or not.

It bears thinking about.

Certainly your prevailing style, and what a particular document says about you, are worth thinking about. If there is an inevitable subtext of this sort, you cannot afford to let it go by default; you need to influence it consciously. Start by considering what you want people to think of you. Take a simple point. You want to be thought of as efficient. Then the style of the document surely says something about this. If it is good, contains everything the reader wants, and certainly if it covers everything it said it would, then a sense of efficiency surely follows.

The same applies to many characteristics: being seen as knowledgeable, experienced, authoritative and so on (a concept covered earlier and worth noting here in context). All such characteristics are worth considering to ascertain exactly how you achieve the effect you want. Such images are cumulative. They build up over time and can assist in the establishment and maintenance of relationships. Whether such is with a colleague or customer, or concerned with establishing with the boss that you are a good person to work with (as well as good at your work), the influence can be powerful.

Similarly you might have in mind a list of characteristics you want actively to avoid seeming to embrace. For example appearing dogmatic, patronizing, inflexible, old-fashioned or whatever in your job might do you little good. Some other characteristics are sometimes to be emphasized, sometimes not. Stubbornness is a good example.

Such images are not created in a word. There is more to appearing honest than writing, 'Let me be completely honest' (which might actually just ring alarm bells!). Your intended profile will come, in part, from specifics such as choice of words, but also from the whole way in which you use language. So it is to more about the use of language that we now move on.

10.4   The use of language

How language is used makes a difference to exactly how a message is received. The importance of using the right word has already been touched on, but the kind of difference we are talking about can be well demonstrated by changing no more than one word. For example, consider the first sentence of this paragraph: 'How language is used makes a difference to exactly how a message is received.' Add one word: '… makes a *big* difference to …'

Now let us see what changing that word 'big' makes: it is surely a little different to say: '… makes a *great* difference …' And there are many alternatives, all with varying meaning: *real, powerful, considerable, vast, special, large, important*. You can doubtless think of more. In the context of what I am actually saying here, *powerful* is a good word. It is not just a question of how you use language, but what you achieve by your use of it.

Note: no business writer should be without both a dictionary and thesaurus beside their desk; the latter is often the more useful.

10.5   Making language work for you

Often, business writing is almost wholly without adjectives. Yet surely one of the first purposes of language is to be *descriptive*. Most writing necessitates the need to paint a picture to some degree at least. Contrast two phrases: 'smooth as silk' and 'sort of shiny'.

The first (used as a slogan by Thai Airways) conjures up a clear and precise picture – or certainly does for anyone who has seen and touched silk. The second might mean almost anything: dead wet fish are sort of shiny, but they are hardly to be compared to the touch of silk. Further, an even more descriptive phrase may be required: what about 'slippery as a freshly buttered ice rink'? Could anyone think this meant other than *really, really* slippery?

The question of expectation of complexity, and cognitive cost, was mentioned earlier (see Subsection 9.4.1 in Chapter 9), and to some extent it does not matter whether something is short or long. Whatever it is, if it makes things effortlessly clear, it is appreciated. And, if it is both descriptive and makes something easier to understand, readers are doubly appreciative.

Clear description may need working at, but the effort is worthwhile. Trainers often ask a meeting venue to group the places in 'a U-shape'. You can put people in a U around a boardroom-style table. But more often it means a U in the sense of an open U, one that gives the trainer the ability to stand *within* the U to work with delegates. Two different layouts, which both demand precise description.

Description is important, but sometimes we want more than that. We want an element of something being descriptive, and also *memorable*. This is achieved in two ways: first, by something that is descriptive yet unusual; second, when it is descriptive and unexpected.

Let us return to the venue theme above for a moment. Do you remember the hotel's sales executive in Chapter 8 (see 8.10.9) who described, as part of an explanation about room layouts, a U-shape as 'putting everyone in the front row'? He is being descriptive and memorable because, while clear, this phrase is also an unusual way of expressing it. Such phrases work well and are worth searching for.

As an example of the second route to being memorable, I will use a description I once put in a report. In summarizing a 'perception survey' (researching the views customers and contacts held of a client organization) I wanted to describe how the majority of people reported. They liked them, were well disposed towards using them, but also found them a little bureaucratic, slow and less efficient and innovative than they would ideally like. I wrote that they were seen as being 'like a comfortable, but threadbare, old sofa, when people wanted them to be like a modern, leather executive chair'.

This is clearly descriptive, but it gained from being not just unusual, but by being really not the kind of phrase that is typically used in business writing. Its being memorable was confirmed, because it rang bells and at subsequent meetings was used by the organization's own people to describe the changes that the report had highlighted as necessary.

There are occasions where this kind of approach works well, not least in ensuring that something about the writer is expressed along the way. Some phrases or passages may draw strength because the reader would never feel it was quite appropriate to put it like that themselves, yet find they like reading it.

Another element you may wish, on occasion, to put into your writing is *emotion*. If you want to seem enthusiastic, interested, surprised – whatever – this must show. A dead, lacklustre style – 'The results were not quite as expected, they showed that ...' – is not the same as one that characterizes what is said with emotion: 'You will be surprised by the results, which showed that ...' Both may be appropriate on occasion, but the latter is sometimes avoided when it could add to the sense and feeling, and there might be occasion to strengthen that: 'the results will amaze'.

Consider this. How often when you are searching for the right phrase do you reject something as either not sufficiently formal (or conventional)? Be honest. Many are on the brink of putting down something that will be memorable or will add power, and then they play safe and opt for something else. It may be adequate, but it fails to impress; and may well then represent a lost opportunity.

Next, we look at some things to avoid.

10.6   Mistakes to avoid

Some things may act to dilute the power of your writing. They may or may not be technically wrong, but they end up reducing your effectiveness and making your objectives less certain to be achieved. We will look at some examples.

10.6.1   Blandness

Watch out! As has been said, this is a regular trap for the business writer, but it is worth emphasizing here. It happens not so much because you *choose* the wrong thing to write, but because you are writing on automatic pilot *without* thought, or at least much thought, for the detail and make no real conscious choice.

What does it mean to say something is:

- '*quite* good' (or bad);
- '*rather* expensive';
- '*very* slow progress'?

What exactly are:

- 'an *attractive* promotion' – as opposed to a profit-generating one, perhaps; and
- 'a *slight* delay' – for a moment or a month?

All these give only a vague impression. Ask yourself exactly what you want to express, and then choose language that does just that.

10.6.2   'Office-speak'

This is another all-too-common component of some business writing, much of it passed on from one person to another without comment or change. It may confuse little, but *adds* little, too, other than an old-fashioned feel. I am thinking of phrases such as:

- 'enclosed *for your perusal*' (even 'enclosed *for your interest*' may be unsuitable – you may need to tell them *why* it should be of interest – or 'enclosed' alone may suffice);
- 'we respectfully acknowledge receipt of' (why not say, 'Thank you for …'?);
- 'in the event that' ('if' is surely better);
- 'very high-speed operation' ('fast', or state just *how* fast);
- 'conceptualized' ('thought').

Avoid such trite approaches like the plague, and work to change the habit of any 'pet' phrases you use all too easily, all too often, and inappropriately.

10.6.3   Language of 'fashion'

Language is changing all the time. New words and phrases enter the language almost daily, often from America and also linked to the use of technology. It is worth watching for the lifecycle of such words because, if you are out of step, then they may fail to do the job you want. I notice three stages:

1. when it is *too early* to use them (when they will either not be understood or seem silly, or even look like a failed attempt at trendiness);
2. when they *work well*;
3. when their use begins to *date* and sound *wrong or inadequate*.

Examples may date too, but let me try. When BBC Radio 4 talks about an 'upcoming' event, then for some people, this is in its early stage and does not sound right at all; forthcoming will suit well for a while longer.

On the other hand, what did we say before we said 'mission statement'? This is certainly a term in current use. Most people in business appreciate its meaning and some have made good use of the thinking that goes into producing one.

What about a word or phrase that is past its best? Let me suggest a common one: 'user-friendly'. When first used it was new, nicely descriptive and quickly began to be useful. Now with no single gadget on the entire planet not so described by its makers, it has become weak to say the least.

10.6.4   Mistakes people hate

Some errors are actually well known to most people, yet they still slip through and there is a category that simply shares the fact that many people find them annoying when they are on the receiving end. A simple example is the word *unique*, which is so often used with an adjective. Unique means that something is like nothing else. Nothing can be *very* unique or *greatly* unique; even the company whose brochure I saw with the words 'very unique' occurring three times in one paragraph does not in fact have a product that is more than just unique even once. Think of similar examples that annoy you and avoid them too.

Others here include the likes of:

- 'Different to' ('different from');
- 'Less than', as in 'less than ten people' ('less' relates to quantity, as in 'less water', whereas 'fewer' would be correct for number: 'fewer than ten people').

Another area for care is with unnecessary apostrophe's (*sic*), which is becoming a modern plague.

10.6.5   Clichés

This is a somewhat difficult one. Any overused phrase can become categorized as a cliché. Yet a phrase such as 'putting the cart before the horse' is not only well known but establishes an instant and precise vision – and can therefore be useful. In a sense, people like to conjure up a familiar image and so such phrases should not *always* be avoided, and reports may not be the place for creative alternatives such as 'spread the butter before the jam'.

10.7   Following the rules

What about *grammar*, *syntax* and *punctuation*? Of course they matter – so does spelling – but spellcheckers largely make up for any inadequacies in that area these days; though you need to cheque (*sic*) carefully, for there are plenty of possibilities for error that a spellchecker would not pick up. But some of the rules are made to be broken and some of the old rules are no longer regarded as rules, certainly not for business writing.

Certain things can jar. Let us take just three examples.

- **Poor punctuation**: Too little is exhausting to read, especially coupled with long sentences. Too much becomes affected-seeming and awkward. Certain rules do matter here, but the simplest guide is probably breathing. We learn to punctuate speech long before we write anything, so in writing all that is really necessary is a conscious inclusion of the pauses. The length of pause and the nature of what is being said indicate the likely solution. In some ways too much is better than not enough.

- **Tautology**: This is unnecessary repetition, of which a classic example is 'I, myself, personally'. This type of thing is to be avoided. Do not 'export overseas', simply 'export'; do not indulge in 'forward planning', simply 'plan'.
- **Oxymoron**: This is a word combination that is contradictory, may sound silly – 'distinctly foggy' – or be a current good way of expressing something: 'deafening silence'. Some sentences can cause similar problems of contradiction: 'I never make predictions, and I never will.'

Other things are still regarded as rules by purists, but work well in business writing and are now in current use. A good example here is the rule stating that you should never begin a sentence with the words 'and' or 'but'. But you can. And it helps produce tighter writing and avoids overlong sentences. But – or, rather, however – it also makes another point; do not overuse this sort of thing.

Another, similar, rule is that sentences cannot end with a preposition. Yet 'He is a person worth talking to' really does sound easier on the ear than 'He is a person with whom it is worth talking.' Winston Churchill is said to have responded to criticism about this with the famous line: 'This is a type of arrant pedantry up with which I will not put.'

Still other rules may be broken only occasionally. Many of us have been brought up never to split infinitives, and it thus comes under the annoyance category most of the time. There are exceptions however: would the most famous one in the world – *Star Trek*'s – 'to boldly go where no man has gone before' – really be better as 'to go boldly …'? I do not think so.

10.8   Personal style

Finally, most people have, or develop, a way of writing that includes things they simply like. Why not, indeed? For example, although the rule books now say they are simply alternatives, some people think that to say 'First …, secondly … and thirdly …' has much more elegance than beginning: 'Firstly …' The reason for this matters less than achieving an effect you feel is right.

It would be a duller world if we all did everything the same way, and writing is no exception. There is no harm in using some things for no better reason than that you like them. It is likely to add variety to your writing, and make it seem distinctively different from that of other people, which may itself be useful.

Certainly, you should always be happy that what you write sounds right. So, to quote the writer Keith Waterhouse, 'If, after all this advice, a sentence still reads awkwardly, then what you have there is an awkward sentence. Demolish it and start again.'

However carefully you strive to write clearly and in a way that creates an impact, there are certain special circumstances that will tax you still more. Prime among these is when a report is a proposal and its message needs setting out persuasively for prospects or customers. Your internal communication may also need to be persuasive at times.

10.9   Summary

Overall, remember to:

- make sure what you write is not only readable, but is designed for its readers;
- put clarity first; understanding is the foundation of good business writing;
- influence the subtext that provides an image of you, and ensure it works as you want;
- make language work for you; be descriptive, be memorable;
- make your writing correct, but make it individual.

Chapter 11
The different forms of written communication

11.1 Write right

Any written business document must stand up to analysis, and its only real test is whether its reader finds it does the job it was intended to do. This means it must have a clear purpose, and that what it says, and how it says it, are understandable; indeed, that it exhibits any other characteristic that it needs to meet its specific intention and such as has been set out in the last chapter.

In this chapter we look at the approach necessary for different types of document. It should be noted at once that there is no one right way to write anything. A variety of word combinations and of style can be appropriate, but there are some things that must be done in particular ways and there are certainly things to avoid. We start with some general thoughts.

11.2 First principles

It is easiest to analyse writing through an example, so let's start with a typical business letter. This is something that many of us have received, usually addressed by name and slipped under the door to greet us as we rise on the last day of a stay in a hotel. The example that follows is a real one, though the originator's name (a city-centre hotel) has been removed.

11.2.1 Example

Dear Guest

We would like to thank you for allowing us to serve you here at the [Name] Hotel and hope that you are enjoying your stay.

Our records show that you are scheduled to depart today, and we wish to point out that our checkout time is 12 noon. Should you be departing on a later flight, please contact our front desk associates who will be happy to assist you with a late checkout. Also, please let us know if you require transport to the airport so that we can reserve one of our luxury Mercedes limousines.

In order to facilitate your checkout for today, we would like to take this opportunity to present you with a copy of your up-dated charges, so that you may review them at your convenience. Should you find any irregularities or have any questions regarding the attached charges, please do not hesitate to contact us.

We wish you a pleasant onward journey today, and hope to have the privilege of welcoming you back to the hotel again in the near future.

Sincerely yours,

[Name]

Front Office Manager

What are we to make of such a letter? It is, necessarily, a standard one used many times each day. It came to my notice when it came under my door and, taking note of the bit about late checkouts – '... will be happy to assist you' – I went to Reception to take advantage. Not only was I told, 'Sorry, we're too full to do that today,' but so were a dozen other people during the ten minutes I stood at the desk. So, the first thing to say is that the letter is so badly expressed that it does more harm than good, causing as much disappointment as satisfaction, because it says clearly that something *will happen* when it should say it *may be* possible.

It is also very old-fashioned with rather pompous-sounding phrases such as 'we wish to point out that' and 'we would like to take this opportunity', when something shorter, more straightforward and businesslike would surely be better. It almost *suggests* that the account may be wrong ('irregularities'), and everything is expressed from an introspective point of view: 'we', 'we' and 'we' again leading into every point. No, it is not good and your own analysis may well run longer. Yet this is very straightforward and hardly technical.

At base, the key problem is perhaps intention. Is the letter designed:

- simply to remind people to pay the bill;
- to make checkout quicker or easier;
- to sell a transport service to the airport;
- to persuade people to come and stay again (and thus presumably give an impression of efficiency and good service); or
- just to say 'Thank you'?

Because it mixes up all of these to some extent, it fails to do justice to any of them. For example, nothing about the checkout procedure is explained, nor are reasons given as to why someone should stay again. Yet this is surely a straightforward letter; perhaps that is why it has been given inadequate thought.

11.3   Letters with specific intention

Often, a document has a very specific purpose, and it must use the way it is written to achieve that aim. There are purposes without number. Here are some that are both important in their own right and make good examples of anything special needing to be executed in a way that reflects its purpose.

Again let's examine an example, this time of a persuasive – sales – letter. The example is a letter to a customer, but similar principles will apply whenever persuasion is necessary.

Sales letters, those specifically designed to be persuasive rather than just administratively efficient, are a key element of written communication with customers.

Whatever their purpose, all must have a clear structure, must use language to make what they say interesting and customer-orientated and aim to be persuasive.

11.3.1 Example

The letter that follows is responding to an enquiry. The intention here must be clear. You cannot write a letter, or write most of it, and then decide how to finish it off and what action to request from the customer. Logically, you must decide what action you want, and then write a letter that is designed to prompt it.

Consider the following, a letter received following a telephone call with a hotel about the possibility of conducting a training seminar there on a quoted date (their quoted cost is omitted).

Dear Mr Forsyth

Following my telephone call with you of yesterday I was delighted to hear of your interest in the [Name] Hotel for a proposed meeting and luncheon some time in the future.

I have pleasure in enclosing for your perusal our banqueting brochure together with the room plan and, as you can see, some of our rooms could prove most ideal for your requirement.

At this stage, I would be more than happy to offer you our delegate rate of [sum] to include the following:

* morning coffee with biscuits
* 3-course luncheon with coffee
* afternoon tea and biscuits
* flip chart, pads and pencils
* room hire and visual aid equipment
* service and tax

and I trust this meets with your approval.

Should you at any time wish to visit our facilities and discuss your particular requirements further, please do not hesitate to contact me but, in the meantime, if you have any queries on the above, I would be very pleased to answer them.

Yours sincerely

Let us consider that for a moment. It is sadly not so untypical in style. Yet, while no doubt well intentioned and polite and containing a certain amount of information, it does not really begin to *sell* in an appropriate manner. Nor does it project a useful image.

Let us look at it again (from the beginning).

- It links to the enquiry but has a weak, formulaic start (and no heading). The enquirer does not want to know about their delight (of course they want the business). Starting with something about the client would be better.
- The enquirer is not running a 'meeting and luncheon' – they explained it was a training session – so this is their terminology, not the client's.
- The event is not 'at some time in the future'. The enquirer quoted a date (this and the point above tell us it is in all likelihood a standard letter).
- Next we have more of their pleasure. They will be more interested in what the brochure will do for them, rather than what their sending it does for the hotel (and, yes, people really do use words like *perusal* in writing, though it seems very old-fashioned to most people – and who would *say* it?).

- 'Banqueting brochure' is jargon – their terminology again (though it may well be useful, and a room plan is useful).
- Do they have a suitable room or not? The words 'some of our rooms could' are unclear.
- The section about costs starts with the words 'At this stage'. But I am sure they do not mean to say, 'we will negotiate later'. The phrase is padding, akin to people who start every sentence with the word 'basically'.
- Most will find the list OK, but is it right to ask if it 'meets with your approval'?
- People who use hotels nearly always want to see something like a meeting room in advance, so the text would be better to assume that and make arranging it straightforward. Also, the writer might better have maintained the initiative and said they would get in touch (they never did, incidentally).
- Suggesting there may be queries is again wrong. Why? Are they suggesting the letter is inadequate? Talking about 'additional information' would be wholly different.
- The cumulative effect of their delight and pleasure – five references, if you include 'I trust this meets with your approval' – is somewhat over the top. They are doing everything but touching their forelock, especially if there are additional things that might be more usefully said instead.

You may find other matters to comment on also: the punctuation is scarce, for example. Certainly the net effect does not stand up to any sort of analysis, bearing in mind its intention is to impress a potential customer.

So, how might it be better done? An alternative (and there is, of course, no such thing as a 'correct' version) is shown below.

Dear

Training seminar: a venue to make your meeting work well
Your training seminar would, I am sure, go well here. Let me explain why. From how you describe the event, you need a businesslike atmosphere, no distractions, all the necessary equipment and everything the venue does to work like clockwork.

   Our [Name] room is among a number regularly successfully used for this kind of meeting. It is currently free on the days you mentioned: 3/4 June. As an example, one package that suits many organizers is:

- morning tea/coffee with biscuits
- 3-course lunch with tea/coffee
- afternoon tea/coffee with biscuits
- pads, pencils and a name card for each participant
- room hire (including the use of a flip chart and OHP)

at a cost of [sum] per head, including service and tax.

   Alternatively, I would be pleased to discuss other options; our main concern is to meet your specific needs and get every detail just right.

   You will almost certainly want to see the room I am suggesting; I will plan to telephone to set up a convenient time for you to come in and have a look. Meantime, our meetings brochure is enclosed (you will see the [Name] room on Page 4). This, and the room plan with it, will enable you to begin to plan how your meeting can work here.

   Thank you for thinking of us; I look forward to speaking with you again soon.

   Yours sincerely

This is much more customer-orientated. It has a heading, it starts with a statement almost any meeting organizer would identify with (and with the word 'your'). Its language is much more businesslike and yet closer to what someone would say, the latter helped by expressions such as 'get every detail just right'. The writer keeps the initiative and sets the scene for follow-up action (while making that sound helpful to the customer and recognizing that they are likely to want to inspect the hotel). Finally, it remains courteous, and putting the (one) thank-you at the end makes it stand out and allows for a much less formulaic beginning. Better, I think.

11.4   Reports

Now consider something about longer documents. First, the greater the length and/or complexity, the more important it is to prepare carefully in order to set clear objectives and focus appropriately on the reader. Beyond that, and details will not be repeated, the length demands a clear structure.

The simplest structure one can imagine is a beginning, a middle and an end. (The exact role of these three parts was covered in the context of presentations and may be worth looking at again in Chapter 6.) Indeed this arrangement is what a *report* must consist of, but the *argument* or case it presents may be somewhat more complex. For example, such may fall naturally into four parts:

1.   setting out the *situation*;
2.   describing the *implications*;
3.   reviewing the *possibilities*;
4.   making a *recommendation*.

The two structures can coexist comfortably; the overriding consideration is logic and organization.

An example helps to spell out the logical way an argument needs to be presented if it is to be got over clearly. Imagine an organization with certain communication problems; a report making suggestions to correct this might follow the following broad sequence.

1.   **The situation**: This might refer to both the quantity and importance of written communication around, and outside, the organization; also to the fact that writing skills were poor, and no standards were in operation; nor had any training ever been done to develop skills or link them to recognized models that would be acceptable around the organization.

2.   **The implications**: These might range from a loss of productivity (because documents took too long to create and constantly had to be referred back to clarify), to inefficiencies or worse resulting from misunderstood communications. It could also include dilution or damage to image because of poor documents circulating outside the organization, perhaps to customers.

3.   **The possibilities**: Here, as with any argument, there might be many possible courses of action, all with their own mix of pros and cons. To continue the example, it might range from limiting report writing to a small core group of

people, to reducing paperwork completely or setting up a training programme and subsequent monitoring system to ensure that some improvement take place.

4. **The recommendation**: Here the 'best' option needs to be set out. Or, in some reports, a number of options must be reviewed from which others can then choose. Recommendations need to be specific: addressing exactly what should be done, by whom, when, alongside such details as cost and logistics.

Before reading on, you might cross-check whether these four section descriptions apply to any such documents you must write, or whether you can think of more pertinent descriptions to divide up any particular category of report with which you may have to deal.

At all stages, generalizations should be avoided. Reports should contain facts, evidence and sufficient 'chapter and verse' for those in receipt of them to see them as an appropriate basis for decision or action.

With the overall shape of the argument clearly in mind, we can look in more detail at the shape of the report itself. The way in which it flows through from the beginning to the end is intended to carry the argument, make it easy to follow and to read, and to make it interesting too, as is necessary, along the way.

Two special features of reports are useful.

1. **Appendix**: An appendix can be used to remove material from the main text and place it at the end of a report (especially detailed or technical information). This allows the main content to be read without distraction and yet also allows readers to check details as they wish.
2. **Executive summary**: This is a summary in the conventional sense but is put at the beginning of a report (rather than at the end, as is a conventional summary) to provide an overview of what follows. Essentially, it acts to say, 'What follows is worth reading.'

Similarly, a long document must look right. It needs adequate space and headings, clear numbering and clear emphasis (this last one delivered by any graphic device, be it italics or boxed paragraphs).

11.5   Formats demanding special approaches

Certain documents need a special approach because of the way they are regarded within the business world and how they are usually experienced. Where such a form exists, it is worth following, not slavishly necessarily, but carefully. A press release is a good example of such a document (though space puts details beyond our brief here).

As a final example, and a plea to approach business writing creatively and sometimes actively avoid conventional language and approaches, consider the following.

Sometimes something special is needed to jolt a reader into action. A second or third communication in a sequence aiming to chase something down is perhaps difficult in that you may feel that your best shot has been sent and you wonder, 'What can I do next?' Such follow-up communications can:

- repeat key issues (but must find a different way to say at least some of their message);
- simply remind (with strong contacts, this may be all that is necessary);
- offer different action (for instance, the first communication to a customer says, 'Buy it'; the second says, 'Let us show you a sample'), or find some more novel way of continuing the dialogue.

The following example is of the last of these. It makes the point that sometimes there is little new left to say, just 'It's me again' – especially if the proposition is good and the only reason for lack of confirmation is timing or distraction rather than that someone is totally unconvinced. In which case the job is to continue to maintain contact, and ultimately to jog them into action, while appearing distinctive or memorable in the process.

After writing a short book for a specialist publisher, I was keen to undertake another topic for them in the same format. Proposing the idea provoked a generally good reaction – but no confirmation. I wrote and telephoned a number of times. Nothing positive materialized – always a delay or a put-off (you may know the feeling!). Finally, while a further message needed to be sent, all the conventional possibilities seemed to be exhausted. Finally, I sent the following brief message:

> **Struggling author**, patient, reliable (non-smoker), seeks commission on business topics. Novel formats preferred, but anything considered within reason. Ideally 100 or so pages, on a topic like sales excellence sounds good; maybe with some illustrations. Delivery of the right quantity of material – on time – guaranteed. Contact me at the above address/telephone number or meet on neutral ground, carrying a copy of *Publishing News* and wearing a carnation.

Despite some hesitation as I wondered whether this was over the top (it was to someone I had only met once), it was sent. Gratifyingly, the confirmation came the following day (and you can now read the result: *The Sales Excellence Pocketbook*, Management Pocketbooks).

Sometimes a slightly less conventional approach – and some seemingly non-business-writing language – works well. You should not reject anything other than the conventional approach; try a little experiment and see what it can do for you.

11.6   Summary

Written communication presents one of the greatest opportunities to shine as a communicator. This is because prevailing standards are not so good; many people find writing a chore and respect anyone who can do it well. At the end of this final chapter about writing, only three points will be emphasized: whatever kind of document you must create, always:

- prepare thoroughly, and set clear objectives, before you 'put pen to paper' as it were;

- write throughout with a focus on the reader;
- use language, consciously, to make what you say powerful.

Finally, let us give the last word to an especially prolific author, Isaac Asimov (who wrote nearly 500 books, mainly science and science fiction). Asked what he would do if told he only had six months to live, he answered simply, 'Type faster.' Clearly, he was someone who enjoyed writing. But his reply is also a good example of the power of language. Think how much his response says about the man and his attitude to life, his work and his readers; and in *just two words*.

Chapter 12

The ubiquitous email: dos and don'ts

Email is one of the quickest ways of communicating with other people, instantly sending, as it does, letters, memos, pictures and sounds from one computer to another via the Internet on a worldwide basis (there are internal networks, too, in larger organizations). The technicalities do not need to concern us here, but the communications implications do – so much so that they deserve their own chapter.

In the working environment, emailing is often used as a substitute for other kinds of communication, reducing the need for face-to-face meetings. This can be useful: it is possible to conduct meetings, correspond with the whole world and use voice and visual contact without leaving your desk.

But the use of email can be overdone, reducing personal contact to the detriment of relationships and collaboration. It is important to have a balance in terms of different forms of contact. Some large organizations have rules to stop any negative effects: no internal emails are to be sent on a Thursday, for instance.

Because emailing is rapid, it brings pressure on the individual responsible for creating the emails to get it right first time in terms of passing a clear message; there is a tendency to 'dash them off', which must be resisted.

The attractiveness of emailing is without doubt its speed. Mail is sent immediately you click the 'Send' button. Your message should be received very quickly after it is sent. The speed of any reply is dependent only on how often someone checks their email inbox and replies. The fast to-and-fro nature of email communication can prompt rapid action and boost efficiency.

12.1   Email versus snail mail

Email can be, indeed usually is, less formal than writing a letter.

Let's say this firmly and upfront. The level of formality must be selected wisely. There are those to whom you may write very informally (incorporating as many abbreviations, grammatical shortcuts, minimal punctuation and bizarre spellings as you wish) *as long as your meaning is clear*. But others (customers, senior colleagues) may resent this or think worse of you for it. Sometimes (usually?) an email must be as well written as any important letter. It is safest to adopt a fairly formal style, and certainly a clear one, and err on the side of more rather than less formality if you are unsure. You have been warned! Proofreading is as important here as with many other documents.

Email's main purpose is not for lengthy communications, but usually for short, direct information-giving or -gathering. Lengthy emails are difficult to read and absorb on screen. For this and other reasons, other means of communication are sometimes better selected (or an email may have a hard copy sent on).

When replying to an email, you don't have to worry about finding the sender's name and address and job title. Replying involves only clicking on a button and their address appears on the top left-hand box of the reply page. It is possible to keep the copy of their sent message on the page, so that you can refer to it when replying, and they can refer to it when reading your reply.

As an example of what is possible, a company located in Wisconsin, USA, emailed its service consultants in Cambridge, UK, about obtaining a specific part for a processing machine. The UK office emailed the manufacturers in Manila, Philippines, for information. They responded by email within minutes. The reply was then transmitted back to the US company. Total time taken: 17 minutes to circumnavigate the world and deliver what was regarded as exceptional service.

12.2   Email – possible disadvantages?

Email is not universally wonderful, for several reasons.

- It is obviously impossible to communicate with someone electronically unless the recipient has a computer set up to receive email.
- Email agreement is just as legally binding as a formal document; treating it otherwise can cause problems.
- If technical problems put your system out of action this can cause problems; and technical backup needs to be in place (it is not a question of *if* it happens, but *when*).
- Most junk email – or 'spam' – is just as irritating as the junk mail that arrives through the letterbox. The responsibility rests with the user and it is sensible to reduce its volume by having, and keeping up to date, software that isolates it.
- Caution should be exercised in opening emails and attachments from unknown recipients, as viruses, Trojan horses and worms can invade the computer system if care is not taken; more of this later.

Already enough has been said about email for other problems to be apparent. People sending personal messages can waste much time in an organization. If this is done on a company heading or format, there may be legal implications too: what happens, for instance, if something is libellous? Thus, organizations need firm policy and guidelines and everyone needs to be disciplined in following the rules.

12.3   Some basic guidelines

As has been said, emails can be more informal than letters but still certain criteria as regards style and content are sensible (again, an organization may set out guidelines).

Given the volume of emails people receive, you are competing for attention and must compose emails that are effective. An email should be:

- **brief** – use plain words;
- **direct** – clear presentation, no ambiguity;
- **logical** – with a clear structure.

Whether emails are being sent internally or externally, as a substitute for a letter or not, it is important to ensure these rules are observed. A clear heading will make its purpose apparent and it may also be helpful to flag any (real!) urgency and say whether, and perhaps when, a reply is sought. Remember that email can, like any communication, have many intentions – to inform, persuade, etc.

Before sending an email, considering the following will help ensure that it is presented effectively.

1. **What is the *objective*, or purpose, of email?** Do you know what you are trying to achieve? Is the email a request for information? Are you circulating standard information? If the email is a quick response to a query, make sure that what you say is correct. If you are unsure, explain that this is an acknowledgement of receipt, and you will come back to them as soon as you can. If you do not know what the objective is, think carefully before sending your communication.

2. **What is the *background* to the issue?** Is the reason for sending the email something that is to do with a problem in a project? Is there an explanation, excuse or apology required? Is it to elicit more information or to provide detailed answers to a query? For an email to be clearly understood, there must be a reason why you are sending it. If you don't know, check before going into 'print'.

3. **Who is the intended *recipient*?** Will it reach them direct, or be read by another person? Email inboxes are not necessarily opened only by the person named in the 'Send to' box. It is possible that colleagues have access to a person's mailbox, for example when someone is sick or on holiday. It is important to bear this in mind when writing a message in case of problems.

4. **What *style* are you using?** How is it being presented? Is the style really informal? Are you replying to a message that was half encrypted with lots of missing capital letters, text-message-style shortened words, emoticons, etc.? If so, that is fine. But think carefully what impression the style of the email gives to someone who is opening a communication from you for the first time.

5. **Choice of *content*.** What is the email saying and is it being clearly communicated without any vagueness and ambiguity? If the email covers complex matters, it may be better to explain that a document follows. It is usually intended for emails to be read quickly, and the content should reflect this.

6. **Is a *conclusion/recommendation/response* required?** If so, is this obvious? It may be clearest to place any request for action at the end of the email. Also, by saying something like, 'It would be helpful if you could bring this information with you when we meet at 4 p.m.,' you give the recipient a clear message that they have until 4 p.m. to complete the task. Finishing off an email with a direct instruction, or repeating the purpose of the message, will leave the reader in no doubt about what your intention is.

7. **What, if any, *attachments* are being sent?** Specify any attachments clearly. If a device is used to 'squash' information together, to compress it – such as a zipping program – it is always helpful to explain which system you use. If the attachments require certain software to open them, explain what is needed. This is particularly important where graphics and images are being sent. Some of these attachments can take ages to download and it is helpful to say so.

Putting yourself across appropriately in an email is important, because it is instant and nonretrievable. As with other written communication, there is no tone of voice, facial expression, posture, body language or gestures to augment your message. As email is a rapid and concise form of communication, the detail matters (see Panel 12.1).

Panel 12.1   Getting the detail right

These are some of the most important points of detail to remember.

- **Format**: Use an appropriate format or house style – this is often available as a template. Make sure it matches the style used in the company's letters and faxes and check what other aspects of layout are expected to conform.
- **Typography/font**: Most companies have a prescribed font and style but others can be chosen from the dropdown list box. This often shows how the font actually looks. You can also select the option to point up text in your email, using devices such as **bold**, <u>underline</u> and *italic*.
- **Subject**: Writer reference, case number or project name. This is just a polite way of ensuring that the recipient can save time by reading what the email refers to. If you are sending an email to someone about a particular matter, it is helpful if they understand immediately what the message is about.
- **Salutation**: Are you on first-name terms? Do you need to write in more formal style because you have not exchanged correspondence before? Do you know the name of the person to whom you are writing, or would it be an impersonal salutation?
- **Punctuation**: Beware ambiguity. A missing comma or no full stop can often cause confusion. It may be 'cool' to lose capitals and miss out dots and dashes, but, if the reader is left puzzled by the meaning, you are less likely to get a useful exchange of information.
- **Line length**: Short sentences and line length make for easier reading on screen. This is explained in more detail further on. Do not use complex sentences or syntax. Short and sweet is best.
- **Paragraphing**: Options are usually available from the dropdown list, including headings, bulleted and numbered lists. A new paragraph should be introduced where there is a change of topic or subject, so that the reader is aware that a new point is being introduced.

Panel 12.1   *Continued*

- **Consistency**: If the email contains numbering, take care. It is extremely irritating if the numbering changes in style or is inconsistent. If you are making a number of points, stick to a) b) c) or 1.i, 1.ii, 1.iii or whatever style or format you prefer.
- **Valediction**: Unlike with a formal letter, you don't have to sign off 'Yours faithfully' or 'Yours sincerely'; however, in some cases it may be appropriate to end with an informal send-off. Many people use 'Kind regards', 'Many thanks' and 'Best wishes', or, more impersonally, 'Yours'.
- **Writer details**: Title, company. With emails it is possible to set up as a 'default' a signature, which appears at the foot of the message. This includes your name and title as well as details of the company you represent.
- **Contact details**: These go with the signature and should include any contact details necessary, such as those that appear on the company letterhead (telephone and fax numbers etc.).
- **Attachments**: As mentioned before, these should be clearly described and mentioned in the text. If they are in a different format, such as PDF files or zipped (compressed, sometimes called 'stuffed'), it is a good idea to ensure beforehand that the recipient's computer is able to receive these files in readable form.

Note: While these are especially important in context of the special nature of email, they overlap with the general principles of what makes any written message effective. These were dealt with earlier.

12.4   Time-wasting emails

It is an intractable problem, but some points are worth noting. Junk emails are a nuisance and can be time-consuming. Emails received from reputable bodies sending legitimate commercial email, as compared with 'illegal spammers', are within the law. The majority of illegal spam emails can be readily identified from the address and/or subject and immediately deleted without being opened. Ninety per cent of spam comes from countries such as the USA and Russia.

Replying to illegal spam will often make things worse. The spammer will know that your email address is valid, will continue to use it and circulate it to other spammers. Anti-spam filters have been mentioned and are incorporated in corporate IT systems as standard. They do not catch everything, but they certainly reduce the volume of spam reaching email in boxes.

You can block some unwanted messages. There may be a number of reasons why emails need to be blocked from particular senders. Junk emails are just one of the main reasons; others include people with whom you no longer wish to correspond. By setting up barriers supplied with your email package, you can block specific email addresses. You can easily remove a sender from the blocked-senders list by selecting the address and clicking the 'Remove' button. When a sender is blocked, their message will be diverted straight into the 'Deleted Items' folder. Do not forget to empty this folder often, otherwise it can become clogged with unwanted messages.

Beware of opening messages from unidentifiable sources, particularly with attachments. These can contain viruses or micro-programs that can access your information and send it to others.

12.5   Digital signatures and other security devices

Several other things should be noted with regard to security.

- *Electronic signatures* are being used more widely as more people send information by email. In addition, it is more important than ever that emails cannot be read by anyone other than your recipient. By using digital IDs or signatures you can ensure that no one is pretending to be you, and sending false or misleading information under your name. Digital IDs in Outlook Express can prove your identity in electronic transactions, rather like producing your driving licence when you need to prove who you are. Digital IDs can be used to encrypt (code) emails to keep the wrong people from reading them. Digital IDs are obtained from independent certification authorities whose websites contain a form that, when completed, contains your personal details and instructions on installing the digital ID. This is used to identify emails and ensure security of your messages.
- *Encryption* is a special way to send sensitive information by email. It is a form of electronic code. One code is used to encrypt the message and another code is used to decrypt it. One key is private and the other is public. The public key is passed to whoever needs to use it, whether they are sending the message (in which case they would use it for encryption) or receiving the message (they would use it for decryption).
- *Records*: Some email systems allow a note to be shown when an email has been sent, received, opened and read by the recipient. This can be important in some time-critical instances.

12.6   Viruses

It is advisable not to open emails that may contain *viruses*. A virus is a small piece of code deliberately buried inside a program. Once the program is run the virus spreads and can damage the data in your files or erase information on the hard drive. Thousands of viruses exist and new ones are being invented every day. *Antivirus software* that can search out and destroy a virus is essential. Such should be updated on a regular basis.

Virus checkers should be run before opening any emails with attachments. Some ISPs are now scanning emails on their way through and will inform the email account holder if any possible problem emails have been detected. You can then make an informed decision as to whether you are going to risk opening up the email or not.

Viruses are distributed over the Internet in a number of different ways, for instance:

- *software downloaded* from the Internet itself may contain a virus; they can be transmitted by an email attachment (it pays to be suspicious of an email attachment from an unknown recipient);
- a *macro virus* is hidden inside a macro in a document, template or add-in; a document with a macro virus can not only infect your computer but also other computers if you pass the document on.

Email files can become very important. Losing them can be a disaster. It is advisable, therefore, to *back up* your files and company data *regularly and often* to safeguard against such risks. If passwords are used in your computer system, consider changing them on a regular basis to stop hackers and avoid misappropriation.

There may be company guidelines about this, indeed about backing up the whole computer; but the responsibility is likely to be personal.

12.7 'Email-speak' – the role of language

We have already touched on the lack of formality of many email messages. But clarity is essential and many messages must look and sound good, so too much informality is a danger. So bear in mind the following.

- **Spellcheck your emails when necessary**: Be aware of easily confused words and use the spellcheck with caution. For example, see how easily a sentence meaning is changed by the substitution of the word 'now' with 'not', and vice versa. As an example of the dangers, consider a letter saying, 'After further consideration I have decided that your request for a salary increase of £10,000 per annum will now be agreed.' Try that sentence again inserting the word *not* instead of *now*.
- **Similarly, use grammar and language checks and such features as the thesaurus**: All help produce an effective message. It is possible to select alternative words or phrases to avoid confusion when using the grammar-check tool.

There is a good deal in the earlier chapters about pure writing skills, so suffice here to say that such things as the following make the essentially simple email unwieldy and less likely to do a good job:

- *overcomplexity* ('from time to time' instead of 'occasionally', 'it is necessary that' instead of 'must');
- *tautology or unnecessary repetition* ('new innovation', 'close proximity' and 'postpone until later');
- *unnecessarily long words* ('acquiesce' instead of 'comply', 'requirement' instead of 'need');
- overlong sentences.

12.8   Jargon, initialisms and acronyms

Emails seem to attract abbreviations. Those containing jargon, text language and acronyms (where initial letters are used to make up another word) are more likely to be confusing. However, because of email's overriding informality, it is a good idea to be familiar with those that are universally used. There are many around and new ones are springing up daily due to the popularity of text-messaging. Here is a selection of some of the more common initialisms and acronyms that you are likely to see in emails:

AFAIK as far as I know
BCNU be seeing you
BTW by the way
CUL8R see you later
FYI for your information
TNX thanks

It would be a good idea to learn these and any others that are commonly used within your company, profession or industry sector. Beware of using them in emails that are being sent externally, where the recipient may not understand them. When using acronyms or any abbreviations, do not assume others will understand; it is courteous to use full terminology in parenthesis afterwards.

(There's a handy little book on the market called *The Total TxtMSg Dictionary* by Andrew John and Stephen Blake (Michael O'Mara Books, 2001), which has a lot of these abbreviations, plus many standard ones and a few unusual emoticons and text 'pictures'.)

12.9   Attachments

Email is made infinitely more useful because documents and files can be attached to emails. Attachments can include word-processed documents, images, sound or video files. It is even possible to email computer programs.

When an attachment is sent, the email program copies the file from where it is located and attaches it to the message. Image files can take some time to upload and download, so it is advisable to keep these to a minimum if speed is of the essence.

It may help to compress files that are being sent as email attachments. This will reduce the upload time while transmitting the information. It also speeds up the download time for you if someone sends you a large file that has already been zipped. WinZip and Stuffit are well-known programs and Microsoft Windows includes a compression tool in its later versions.

The advantages of sending documents and files as attachments are the speed and efficiency of communications. The recipient of the documents will be able to keep these on file and can edit, return or forward them as necessary.

If security is an issue, an attachment should be sent as a PDF (Portable Document Format) file. This system gives the compiler the option of setting security restrictions to prevent copying, editing or printing, or any combination of these, if desired. A password can be set to prevent the document from even being viewed except by the intended recipient(s). These settings are flexible and, depending on how the security is set, a PDF could, say, be printed off but not amended. This is very safe and secure for sensitive material.

12.10 Hyperlinks

Inserting hyperlinks into email messages is particularly useful when sending information to people. If you want to alert your recipient to a website or particular web page within a site, simply insert the *hyperlink* into the email message. The recipient then clicks on the link and opens the web page. Remember, though, that this takes time, and some people may not bother; information sent in this way to a customer as a part of a total message might thus never be seen and dilute the whole effect. It is also possible that some recipients may have set their email programs not to accept HTML (hypertext markup language), so any web address in your email will be in plain-text form and will need to be copied and pasted into the browser for the website to be accessed. Again, this takes time – even more time than clicking on a hyperlink. It is, however, a good way to enhance technical information.

12.11 Staying organized

Emails, because of their popularity and versatility, currently threaten to obliterate all other forms of communication. It is vital to stay on top of them, so:

- *clear* the inbox every day;
- *categorize* items that are time-critical and items that will require work later;
- *delete* any emails that are irrelevant or unimportant;
- unsubscribing from email lists assists *clearing the clutter* in your in box;
- *copy yourself* (CC) or blind-copy (BCC) yourself a message when responding to emails arranging a meeting or promising a response or sending information;
- *ration* the reading of emails to, say, three times a day: early morning, midday and end of day; reading messages as soon as they flash on the screen causes severe interruptions;
- *delete messages* that you've dealt with and empty the deleted folder frequently (unless it is of sufficient importance to archive);
- when replying to messages where your reply is integrated into the original text, make sure that the responses you insert are in a different colour to draw attention to revisions and insertions.

Make a list of things that make your emailing more effective and make these points a habit. For instance, here are some don'ts and a few more dos.

- **Don't** send emails just because they are easy.
- **Don't** enter text IN CAPITAL LETTERS. It is taken as shouting.
- **Don't** use them as a substitute for properly delegating a task to another.

- **Don't** send them to discharge yourself of responsibility.
- **Don't** put something in an email that is confidential – it can be abused.
- **Don't** forward someone's email without their permission.
- **Don't** assume your recipient wanted it and is desperate to receive it.
- **Do** think, and use the 'send later' button to inject some thinking time.
- **Do** be precise – to eliminate follow-up phone calls.
- **Do** reply promptly. Because email is quick, a reply is generally expected.
- **Do** be polite and friendly, but never assume familiarity with jargon.
- **Do** keep attachments to a minimum.
- **Do** avoid gobbledegook.

12.12   Summary

Overall, the message here is simple. Use technology where it benefits you, but do so carefully – recognize any downsides and make sure that attention to detail makes what you do effective.

With regard to email, remember that you can tell a lot about a person from their 'email exterior'. An email can provide a window to someone's status in the workplace, work habits, stress levels and even their personality.

Managers who send emails tend to use 'higher-status' techniques. They contain a greater level of formality and tone, and lack the detail of a lower-level member of staff, and you will rarely see cheesy quotes, smiley faces (emoticons) or joke mails.

Emails are such a valuable communication tool for today's managers, but, if abused or used carelessly, they can cause trouble. In summary, here are nine basic tips for better email technique.

1. **Use email as one channel of communication, but not the only one**: This is important, do not be lazy just because it is fast and easy. Emails can document discussions and send high-impact messages around the world at the click of a mouse. But they can also mislead managers into thinking they can communicate with large groups of people solely through regular group emails. Use email widely but not as a management tool. It is not possible to reach everyone, and the 'impersonal' non-direct contact means that people feel can feel slighted by the loss of the personal touch.
2. **It pays to keep it short and sweet**: Emails that are longer than a full screen tend not to be read straightaway. They get left till later and often not until the end of the day or the next morning. It is important to judge when it is right to put down the mouse and seek the person out for a face-to-face encounter, or pick up the phone and speak to them.
3. **Message clearly – cut out the codes**: Email requires clarity of purpose. Be sure your message comes across without any doubt or misunderstanding. Also, it is important to be sure to whom your message needs to be addressed, and who needs a copy for information. In terms of actions and priorities, use lists or bullet points for clarity. Response buttons (or similar) should be used if you need to see who has received and read your message.

4.   **Encourage open communication when using email**: Do this by requesting that people respond with questions or queries if they wish. It shows that you are concerned and available to help.

5.   **Do not use emails to get mad with people**: Far better to save anger for face-to-face encounters (where facial expression and body language can be used to great effect) or over the phone, where tone of voice can say a lot. Sarcasm, irony, criticism or venom is not appropriate when sending emails. They often come over far more harshly than intended.

6.   **Humour should be used with caution**: By all means use wit and humour to lighten a heavy atmosphere, but emoticons, smiley faces and joke mails are not usually appropriate in the work environment. If being facetious is usual for you, it may make it more difficult to strike a serious note when you need to. Some companies ban joke emails; they are too risky. Too many joke emails erode your attempts to send serious ones.

7.   **Suspend reaction – use the five-minute rule**: It is often wise to delay sending a hastily written email for five minutes (or more!) before pressing the 'send' button. If you are angry or upset when you write something, it is a good idea to take a break, go for a walk or do something else, before writing. Otherwise, if you do write the message immediately, once you have cooled off take a moment to review it before sending it out.

8.   **Set aside time to deal with emails**: Because of the growing importance of emails in terms of mode of communication, you need to make time to deal with them; if this demands reconfiguring your working day, so be it.

9.   **Take advantage of tools such as spellcheck and thesaurus**: To avoid errors and complicated sentences, use the tools provided to ensure clarity of communication. If you are unsure whether something works, check – or ask a friend or colleague.

10.   **Next time you send an email, double-check it against the principles set out in this chapter to make sure it is effective**: Also do *not* send copies to all and sundry – this is a prime cause of the fact that most people complain that they get too many emails and that many are unnecessary.

Chapter 13
Dealing with numbers

A common element of presenting technical information, and one that makes a good example, is that of presenting numbers, including financial figures.

13.1   The nature of numbers and number 'blindness'

Numbers can confuse or clarify. The job is normally to make sure they do not confuse, when they should make things clear – though perhaps it should be acknowledged that sometimes numbers are thrown around precisely in order to confuse. For example, someone in a meeting might rattle through a mass of disparate costings in the hope that just how expensive a plan is will not be dwelt upon. Similarly, the complexity of figures may be used on a grander scale: in marketing, people talk about 'confusion pricing' – a pricing structure of such complexity that it makes it difficult for a customer to undertake comparison with competitors (mobile-phone tariffs are an example of this with which many people are familiar).

That said, the concentration in this chapter is on the positive, using numbers effectively in the course of meetings and more routine corporate communication. This is important because many people:

- assume numbers will confuse them, lack skills in numeracy (finding anything from percentages to break-even analysis difficult) and, because they switch off to figures, they need to be motivated to appreciate them;
- have a parochial attitude to figures, e.g. they can take in things on the scale of their own bank balance, but corporate figures confuse by their sheer size;
- are overwhelmed by the sheer volume of figures (imagine the profusion of figures spilled out by many a computer program).

Because of this, particular approaches are necessary.

13.2   Action to help

The presentation of figures therefore generally needs to be well considered if it is to enhance a message. The first principles are these.

- *Select* what information is presented, focusing on key information and leaving out anything that is unnecessary. This can mean, for instance, that information needs to be tailored; the detailed chart included in a report may be inappropriate to use for other purposes and must be abbreviated. Many slides used as visual aids prove overcomplicated if simply taken from a page in a document, and such seems especially prevalent in items involving figures

- *Separate* information, for example into an *appendix* to a report, so that the main message includes only key figures, and the overall flow of the case is maintained while more details can be accessed if required
- *Separate also information from the calculations that arrive at it.* This can be done using appendices or by such devices as boxed paragraphs in a report.
- *Select the appropriate accuracy* as you present figures. Sometimes accuracy helps understanding, or is simply important, while on other occasions it can confuse, and ballpark figures suit better.
- *Repeat* – repetition helps get any message across. With numbers, natural repetition – for instance, going through them orally and issuing something in writing as well – can make all the difference.
- *Proofread* – numbers must be checked perhaps even more carefully than words in written material; remember that one figure wrongly typed might change things radically, and for the worse.

Sales figures may be up, but there are a variety of ways to describe this, as we can see below.

- *Sales are up.* No detail might be necessary.
- *Sales are up about 10 per cent.* A broad estimate may be fine.
- *Sales are up 10.25 per cent.* The precise figure may be important; and note that it is nonsense to say, as is often heard, 'Sales are up about 10.25%' – the word *about* goes only with round figures and estimates or forecasts.
- *Sales are up about £10,000.* The financial numbers may be more important than the percentage (and can be presented with the same different emphasis as just described for percentages). In addition, what the figures refer to must be made clear. For example, 'Sales of Product X are up 10.25% for the period January–June 2005.' Language can, of course, change all such statements – 'Sales are up substantially' – maybe, as here, just by adding one word.

13.3   Methods of presenting numbers clearly

There are two main presentational ways of ensuring clarity.

13.3.1   Graphs and charts

It is an old saying that a picture is worth a thousand words; this principle relates directly to numbers of all sorts. A graph can convey an overall picture, one immediately understood – sometimes literally at a glance.

Often, two graphs work better than one more complex one, and of course they must be set out to maximize the visibility of the information they display. Thus they need to:

- be an appropriate size;
- use different colours when possible and when this helps – carefully chosen colours, too, picked to contrast one against the other;

- be suitably annotated, with thought being given to what text appears on the graph itself and what is separate (in a key at the foot of the page, perhaps);
- work effectively, or be adapted so to do, if they are to be used as visual aids, when legibility is doubly important.

Various kinds of chart and graph can be used. These include the following.

13.3.1.1   Tables

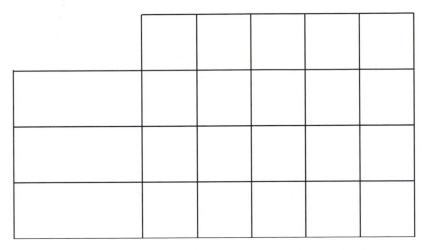

This term encompasses anything that sets out figures in columns; they can be of varying degrees of complexity.

13.3.1.2   Bar charts

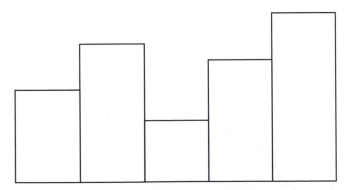

The example makes the effect of this clear; scale can again be varied for emphasis.

13.3.1.3   Pie charts

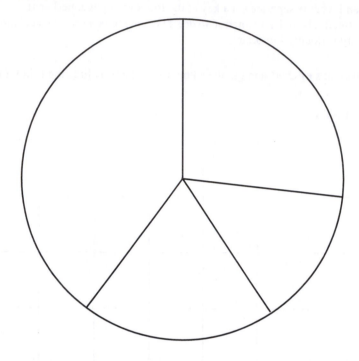

This is an especially visual device and can make many things much clearer than, say, a table and certainly a description.

13.3.1.4   Graphs

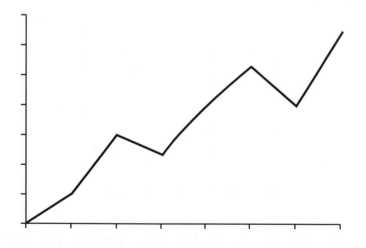

These are useful to show the differences of quantities varying over time. Care must be taken to select scales that give the picture you want (there can be a lot of trickery with this sort of graph – something to bear in mind when you are interpreting them rather than showing them).

13.3.1.5 Project timetables

A	Phase 1	2	3	4
B	1	2	3	
C	1	2	3	4	5

Time

A device to help people visualize the timescale of projects with multiple and overlapping stages.

13.3.1.6 Flowcharts

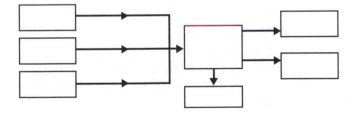

These seem more complex and allow the presentation of a more complex picture – useful where interrelationships are concerned.

All these devices benefit from being kept as simple as possible. Complexity, which includes trying to demonstrate too many different things (such as overall sales, sales by product, profitability and …but you take the point), can quickly drown out clarity.

Note, too, that sometimes a compromise is necessary here between the 'perfection' of a lovingly created graph and the time (and sometimes cost) of producing it. Despite this, the greatest danger is using few, or poorly executed, devices when the information they present makes their use necessary. It can literally be true to say that one graph of some sort added to, say, a presentation can swing an argument or get agreement to a case.

The power of such devices can be considerable. One example of minimal information in graphic form is something I use on presentations-skills courses. By linking an empty pie chart to questions, and adding information as discussion proceeds, first thinking can be stimulated, then information relayed, which will have a good chance of being retained.

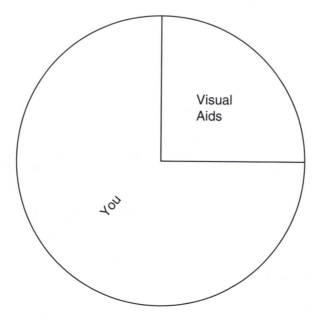

The two segments above get people considering the contribution of what someone says in a presentation and how they express it in relation to the amount that comes over through any slides used. This is relevant to this topic and, in case you wonder, the contribution of the presenter should be larger, as normally slides *support* the presentation rather than lead.

13.4 Language

Certain factors are important here.

- **Pace**: In oral communication you may need to slow down a bit when dealing with numbers and build in more, and longer, pauses.
- **Signposting**: This can focus attention and prompt concentration: 'Let's look at this carefully – it can be confusing and the details are important.'
- **Checks**: Again, in conversation, an adequate number of instances of 'Is that clear?' and similar remarks or questions can help ensure people keep up.

- **Articulation**: Speak clearly. Numbers can be confused and you do not want to have someone thinking you offered, say, a 15 per cent discount, when it was 5 per cent.
- **Precision**: Exactly the right accompanying word can make a difference, ensuring that figures are taken exactly how they should be. For example, 'Note that this figure is an *estimate*' or 'This is the position *today*.'
- **Nonsense**: Some statements using numbers can be, in effect, nonsense. For instance, you often hear advertisements on the television saying things like, 'Now more than 70 per cent ...' High percentages of something good may automatically sound good, but pose the question, 'Compared with what?' It may be a competitor, or an earlier version of the product; or it may just be unclear.

Another element that can help is that of making comparisons. This can be done in various ways. For example, many numbers, not least financial ones, involve comparisons with other periods of time ('This is more than last year') and with other elements ('Administration costs are down, though customer-service ratings are up').

Similarly, comparisons can be made simply to aid description. This is especially helpful if figures are very large or otherwise beyond the day-to-day experience of those for whom the numbers are being laid out. One aspect of this is simply in the words, describing what you want people to take on board as a serious shortfall in revenue, say, as 'akin to the national debt of most of South America'. The exaggeration is extreme and everyone recognizes that this is done only to highlight the importance, and size, of the figure being discussed. Alternatively, you may select a comparison that is accurate and an important detail of what is being described. Thus an office extension might be described as 'the size of a tennis court', when even an accurate number of square metres (and this may be there too) might fail to create an easy-to-grasp and accurate picture.

This links to what is called *amortizing*, which means to spread a figure expressing it as smaller units that are divisions of the whole. For example, an annual cost of say £1,680 can be described as '£140 per month', 'only £140 per month' or 'not even £150 per month'.

As an example, consider a short case. A travel agent is essentially a service and people business. In one particular firm, with a chain of some 30 retail outlets across several counties, competition meant the business was lagging targets. Good market conditions meant it was something that a more active, sales-oriented approach could potentially cure. Initially, management's approach to the problem was to draw attention to it at every level. Memos were circulated to all staff. The figures – the sales revenue planned for the business, the amount coming from holidays, flights, etc. – were substantial ones; even the shortfall was some hundreds of thousands of pounds.

The result? Well, certainly the sales graph did not rise. But, equally certainly, morale dropped. People went from feeling they worked for a successful organization to thinking it was – at worst – foundering; and feeling that the fault was being laid at their door. The figures meant little to the kind of young people who staffed the counters – they were just unimaginably large numbers to which they were wholly unable to relate personally.

With a sales conference coming, a different strategy was planned. The large shortfall was amortized and presented as a series of smaller figures – one per branch. These 'catch-up' figures were linked to what needed to be sold, in addition to normal business, in order to catch up and hit the target. It amounted to just two additional holidays (Mum, Dad and 2.2 children) per branch, per week. This was something staff could easily relate to – and which they felt they could actually achieve. Individual targets, ongoing communication to report progress and some prizes for branches hitting and besting these targets in a number of ways completed the picture.

The result this time? The numbers slowly climbed. The gap closed. Motivation increased with success in sight. And a difficult year ended with the company hitting the original planned targets – and motivation returned, continuing to run high as a real feeling of achievement was felt.

The key here was one of communication. The numbers and the difficulty of hitting them did not change. The perception of the problem, however, was made manageable, personal and – above all – to seem achievable. The results then showed that success was possible. All that was necessary was to present the figures in the right way – one that reflected the realities of the situation.

Finally, as a thought linked to language, consider two further factors.

- In oral communication the tone of voice helps get things over – do you, for instance, want a '10 per cent increase' to *sound* good or bad? Make sure it comes over as you wish, and perhaps pause on it to let it sink in.
- In written form, numbers, and the words that accompany them, can be presented with different emphasis too. In **bold type**, or *italics* perhaps, or in a larger typeface. There is also a difference between writing '10%', '10 per cent', and 'ten per cent'; and more variations are possible.

13.5   Summary

Numbers can either enhance communication or they can confuse and, at worst, result in the whole communication falling on stony ground. The key to success is to:

- recognize the *role and importance* of numbers within the message;
- recognize the *difficulty* some people have with numbers;
- choose a *method of communication* that highlights them appropriately, both literally (e.g. bold type) or by appropriate emphasis;
- *illustrate* figures wherever possible, for instance with graphs;
- be *especially clear* in what is said (or written) – use powerful description and take time for matters to be understood.

Afterword

We cannot ensure success, but we can deserve it.

John Adams (US president)

There is an old saying that there is no such thing as a minor detail, that all details are major. This is certainly true of communications. Success, or failure, may occur – or be made more or less likely – by small changes and, if these pages have demonstrated anything, then it is surely that.

To return to a danger identified early on, the greatest likelihood of communications failure comes not from major errors, but rather from inattention, lack of preparation and lack of thought, and an assumption that there will be no problem and that you can wing it. Rarely is this so. Communication, especially of anything inherently posing a problem (and technicalities may well be just that), needs working at. But doing so is eminently worthwhile.

Good, powerful communication achieves the following.

- It *makes things happen*. It prompts discussion, consideration and decision, and may help you influence events and get your own way.
- It *impresses*. How you communicate is a key part of your profile, and, if this is positive, it too helps you achieve things both in your work and in your career.
- It must be *made to work*. The penalties of failure are many and can be serious – but the opportunities, what was described in Chapter 6 (6.2) as 'an open goal', are considerable, and often very considerable. No one – however much the focus of their job is on other, more technical matters – should ignore this area. It is a key to many successes.

Afterword

Index